KT-392-290

Cornwall College
140671

Monitoring Butterflies for Ecology and Conservation

Conservation Biology Series

Series Editor
F.B. Goldsmith

Ecology and Conservation Unit, Department of Biology, University College London, Gower Street, London, WC1E 6BT

In the last twenty years **conservation** has been recognized as one of the most important of all human goals and activities. Since the United Nations Conference on Environment and Development in Rio in June 1992, **biodiversity** has been recognized as a major topic within nature conservation, and each participating country is to prepare its biodiversity strategy. Those scientists preparing these strategies recognise **monitoring** as an essential part of any such strategy. Chapman & Hall have been prominent in publishing key works on monitoring and biodiversity, and with this new series aim to cover subjects such as conservation management, conservation issues, evaluation of wildlife and biodiversity.

The series will aim for texts that are scientific and authoratative and will present the reader with precise, reliable and succinct information. Each volume will be scientifically based, fully referenced and attractively illustrated. It will be readable and appealing to both advanced students and active members of conservation organizations such as the RSPB and County Wildlife Trusts.

Books for the series are currently being commissioned and those interested to contribute, or who wish to know more about the series, are invited to contact the editor or Chapman & Hall.

Other relevant books

Monitoring for Conservation and Ecology
F. B. Goldsmith (editor)

Global Biodiversity
World Conservation Monitoring Centre

Monitoring Butterflies for Ecology and Conservation

The British Butterfly Monitoring Scheme

E. Pollard
Formerly of the
Institute of Terrestrial Ecology
Monks Wood
UK

T.J. Yates
Institute of Terrestrial Ecology
Monks Wood
UK

Published in association with the Institute of Terrestrial Ecology (Natural Environment Research Council) and the Joint Nature Conservation Committee.

Published by Chapman & Hall, 2–6 Boundary Row, London SE1 8HN

Chapman & Hall, 2–6 Boundary Row, London SE1 8HN, UK

Blackie Academic & Professional, Wester Cleddens Road, Bishopbriggs, Glasgow G64 2NZ, UK

Chapman & Hall Inc., 29 West 35th Street, New York NY10001, USA

Chapman & Hall Japan, Thomson Publishing Japan, Hirakawacho Nemoto Building, 6F, 1-7-11 Hirakawa-cho, Chiyoda-ku, Tokyo 102, Japan

Chapman & Hall Australia, Thomas Nelson Australia, 102 Dodds Street, South Melbourne, Victoria 3205, Australia

Chapman & Hall India, R. Seshadri, 32 Second Main Road, CIT East, Madras 600 035, India

First edition 1993

© 1993 E. Pollard and T.J. Yates

Typeset in 10/12 pt Sabon by Falcon Graphic Art, Surrey

Printed in Great Britain by St Edmundsbury Press Limited, Bury St Edmunds

ISBN 0 412 40220 3

Apart from any fair dealing for the purposes of research or private study, or criticism or review, as permitted under the UK Copyright Designs and Patents Act, 1988, this publication may not be reproduced, stored, or transmitted, in any form or by any means, without the prior permission in writing of the publishers, or in the case of reprographic reproduction only in accordance with the terms of the licences issued by the Copyright Licensing Agency in the UK, or in accordance with the terms of licences issued by the appropriate Reproduction Rights Organization outside the UK. Enquiries concerning reproduction outside the terms stated here should be sent to the publishers at the London address printed on this page.

The publisher makes no representation, express or implied, with regard to the accuracy of the information contained in this book and cannot accept any legal responsibility or liability for any errors or omissions that may be made.

A catalogue record for this book is available from the British Library

Library of Congress Cataloging-in-Publication data available

Printed on permanent acid-free text paper, manufactured in accordance with the proposed ANSI/NISO Z 39.48-199X and ANSI Z 39.48-1984

Contents

Foreword

This book describes one of the success stories of European ecology and conservation of the past two decades. Although essentially an account of the British Butterfly Monitoring Scheme, its scope ranges widely across the whole field of butterfly ecology, reflecting the diverse aspects of biology that the scheme has helped to elucidate.

It began very modestly twenty years ago. Four of us gathered in Ernie Pollard's office at Monks Wood, with no greater ambition than to devise a quick and simple method for recording changes in butterfly numbers on the adjoining nature reserve. We came up with a method based on transects. It was not particularly original, but had never caught on for monitoring butterflies, being deemed unreliable because sightings of adults were thought to vary too much with the weather.

The new method differed in two crucial ways: we strictly defined the weather conditions under which recording was permitted, and Ernie Pollard tested the transect counts to see if they actually did reflect real changes in butterfly numbers. He found that they did, and indeed were much more accurate than any of us had dared hope. It was thence a short step to set up new transects on different sites near Monks Wood, and to test the method in grassland and other open habitats, with the ultimate dream of establishing a truly national scheme of butterfly monitoring, akin to the BTO's census of breeding birds.

The beauty of the scheme is that its method is sufficiently simple – and fun – for it to appeal to many amateur naturalists who enjoy having an extra purpose to their walks in the countryside. Indeed this became so popular that naturalists were soon volunteering to make 'Pollard Walks' (as they are affectionately known) on far more sites than could be accommodated within the national scheme. Wardens, too, found that by recording the changing numbers of butterflies in different parts of their reserves, they gained great insight about the needs and local distribution of particular species.

One result of this considerable effort is a unique annual measurement of how different species in one group of insects are changing in numbers at sites throughout the British Isles. This is of inestimable value to conservationists, who can now detect long-term declines (or increases) long before these become apparent through traditional mapping schemes as extinctions (or spreads) in a region. It is quickly clear, too, if the populations on a particular reserve are changing out of synchrony with those on neighbouring sites; when this occurs, it is often in response to some obvious act of management, which can be encouraged or avoided in future, depending on the desired effect.

Another bonus has been the amount of new scientific information that the scheme has generated. It is one of those rare ecological goldmines that produces important data for researchers of almost every aspect of butterfly ecology, as a glance through the chapter headings will indicate. A similar scheme has recently been started in The Netherlands, and it is to be hoped that this excellent book will encourage the spread of such schemes throughout Europe and the temperate world.

J.A. Thomas
Principal Scientific Officer
Institute of Terrestrial Ecology

Preface

There are some 22 000 species of insects in Britain. Compared with most other countries in Europe this is a small number. Of the British species, about 2500 are Lepidoptera and, of these, only 58 are resident or common migrant butterflies. No British butterfly species is restricted to this country and most are commoner in other parts of Europe.

In spite of its relative poverty, the British butterfly fauna is the most thoroughly studied in the world. The reasons for this lie partly in our social history, as the study of natural history was popular amongst the wealthier classes in the eighteenth and nineteenth centuries. More recently, the increasing rarity of many butterflies in Britain has led to a substantial amount of research, to gain the knowledge necessary for their conservation.

Although British butterflies are so well studied compared with butterflies elsewhere, and compared with other insects (apart from a few pest species) everywhere, this certainly does not mean that we know all that there is to know about them. Curiously, knowledge of the natural history and ecology of butterfly species tends, with the notable exception of crop pests, to be inversely related to their rarity. Probably the most thorough studies of British species are of the large blue butterfly, by J.A. Thomas, first during its period of decline to extinction and subsequently after its reintroduction from Sweden, and of the endangered heath fritillary, by M.S. Warren. In contrast, there have been very few detailed studies of common butterflies such as the small heath, which is found in grassland in almost every corner of the country.

In the last 20 years amateur and professional lepidopterists have come together in two schemes for the study of butterflies. The first of these to start, in 1967, was the Lepidoptera Recording Scheme pioneered by John Heath. This scheme was organized by the Biological Records Centre at Monks Wood (ITE) in Cambridgeshire. Records were received from some 2000 recorders and results for butterflies up to 1982 were summarized in the *Atlas of Butterflies in Britain and Ireland* (Heath *et al.*, 1984).

The second scheme to start, in 1976, was the Butterfly Monitoring Scheme, with which we are concerned in this book. This scheme also began at Monks Wood, initially independently of the Biological Records Centre but now incorporated within it and to a degree complementing the recording scheme. The monitoring scheme began with a count around the rides of Monks Wood by J.A. Thomas and E. Pollard in May 1973. Four peacock butterflies, one speckled wood and one green-veined white were seen. Since then many thousands of miles have been covered by recorders in the scheme and tens of thousands of butterflies recorded. Several similar schemes have begun in individual counties in Britain and one in The Netherlands is now in its third year. This book describes some of the information that has been obtained from the British Butterfly Monitoring Scheme and places it in the context of current knowledge of the ecology of butterflies and their conservation.

There is a tendency for research scientists to think of the objects of their study, in this case butterflies, solely in terms of data. From time to time, we need to pause and recognize that, for example, the red admiral that has just been recorded is a thing of beauty, with intricate patterns of behaviour; the particular individual may have arrived after a journey from the Mediterranean or beyond. We hope that we have conveyed some of the fascination of butterflies in addition to presenting a wide range of information about them.

The aim has been to present the information in a form which is accessible to both scientists and those with a more general interest in butterflies. It is important that the statistical basis for the interpretation of results is presented, but this is given as briefly as possible and usually in the form of footnotes.

Much of the original draft of this book was written during 1991, when results were available up to and including 1990. Where possible the figures and tables have been brought up to date with the inclusion of results from 1991. Latin names of butterflies are given in Appendix A and of plants in Appendix C. Nomenclature of butterflies follows Thomas and Lewington (1991).

We are aware that the work we describe is the result of the efforts of many people. We very much appreciate the enthusiasm and dedication of the recorders, on whom the entire project has depended. The main recorders at each site are listed in Appendix B. In addition to the authors, Marney Hall and Joan Welch have been closely involved in the running of the Butterfly Monitoring Scheme and we gratefully recognize their major contributions. Many people have supported the monitoring scheme in other ways and amongst these we thank in particular S. Ball, J.P. Dempster, D.O. Elias, P.T. Harding, K.H. Lakhani, R. Leverton, I. McLean, G.J. Moller, M.G. Morris, D. Moss, M.J. Skelton, A.E. Stubbs, J.A. Thomas and M.S. Warren. We would also like to thank G. Swindlehurst, who drew the figures for the book, and B.N.K. Davis for help with the section on St. Osyth. The

Butterfly Monitoring Scheme has been supported financially by the Nature Conservancy Council, the Joint Nature Conservation Committee, and the Institute of Terrestrial Ecology (Natural Environment Research Council).

Figures 4.2–4.4, 5.1, 13.1, 13.5, 13.9, 13.13, 13.17 and 13.21 are modified from Pollard *et al.* (1986) and are published by kind permission of English Nature. The drawings of butterflies in Chapters 9, 11 and 12, by D. Redpath, are based on published photographs from several sources.

The preparation of a book, with a planned structure, may give the impression that the studies it describes developed in a similar planned manner. This is true of the Butterfly Monitoring Scheme to a degree, but we have been frequently surprised by the range of information that monitoring has provided and often by the results themselves. Monitoring is usually regarded as inferior to experimental research, in that it provides data but not understanding. This can happen, but clues to underlying mechanisms are often present in the data. These clues may need to be explored further by more detailed experimental studies; nevertheless, through the years of monitoring we feel that we made a contribution to fundamental knowledge of the biology of butterflies. We hope that the reader will share this view.

This book is dedicated to
the recorders who made
it possible

1

The current status of British butterflies

1.1 INTRODUCTION

When we first considered the possibility of monitoring changes in butterfly numbers, in the early 1970s, we were not fully aware of the severity of the contraction of the ranges of many butterflies. A major stimulus for the early monitoring trials was the knowledge that many claims and counter-claims had been made in the 1960s, when the effects of organochlorine insecticides on wildlife were causing concern, as to whether there had been a decline in the abundance of the commoner butterflies. This argument was unresolved, for the simple reason that there were no data. Many people had youthful memories of clouds of butterflies in summer meadows, but such memories are highly subjective and almost valueless as evidence. An objective method of recording butterfly abundance was required.

As the work on monitoring methods progressed and eventually the national scheme got under way, we became increasingly aware that there had been dramatic contractions in the ranges of many of the rarer butterflies, especially in the last few decades. When the records of the recording scheme for Lepidoptera, run by the Biological Records Centre, were summarized (Heath *et al.*, 1984), 20 (35%) of the 57 species resident in 1900 were considered to have experienced major range contractions. These 20 species include two extinctions during this century (the black-veined white and the large blue). Some of these contractions of range had been in progress over much of the century, but the sharpest declines appear to have been since the 1950s. In this chapter, the aim is to give a concise account of the more recent changes in the status of British butterflies, to set the context in which the Butterfly Monitoring Scheme has operated.

1.2 CONTRACTIONS OF RANGE

There is now a fairly general consensus (Heath *et al.*, 1984; Thomas, 1984, 1991; various authors in Emmet and Heath, 1989; Warren, 1993) that the primary reason for the decline of the rarer butterflies has been the destruction and alteration of their habitats. The declines have been most severe in the east of England, where agriculture is most intensive and pressure on semi-natural areas greatest. Of the butterflies listed in Table 1.1, the decline of at least five, the chequered skipper, pearl-bordered fritillary, high brown fritillary, heath fritillary and Duke of Burgundy, is thought to be associated, at least in part, with the reduction in area of coppiced woodland. Four species, the silver-spotted skipper, Adonis blue, marbled white, and the Duke of Burgundy again, have declined as the amount of chalk downland has been reduced, mainly through ploughing. Of the downland species, the marbled white has declined much less than the others and its relative success can also be related to its habitat requirements; it flourishes in the lightly grazed grasslands on many of the downlands that have survived.

For most of the other species listed as contracting in range, it is also fairly easy to relate their declines to the loss of a particular biotope or biotopes. However, it may be misleading to conclude that the reasons for the decline of all these species are clear-cut, well known and invariably related to known changes in biotopes. Even for those species for which the association with a particular endangered biotope is clear, it is quite likely that other factors, especially weather, have played some role in their changes of status.

The decline of butterflies is by no means restricted to Britain. At a recent symposium on European butterflies in The Netherlands, similar losses and range contractions were reported in several countries, including Italy, France and Switzerland. Declines were greatest in The Netherlands, where the land is even more intensively used than in Britain. van Swaay (1990) reports that of 71 species resident in The Netherlands at the beginning of the century, 15 are now extinct. Further south in Europe the position is at present less serious, but the trends are similar. Pavlicek-van Beek *et al.* (1992) provides an outline of the current situation in Europe.

1.3 MATRIX AND 'ISLAND' BUTTERFLIES

One consequence of the increasing rarity of many butterfly species in Britain is that a significant number now have a substantial proportion of surviving populations on nature reserves or other conserved areas. More generally, it is possible (Table 1.1) to divide the British butterflies into those of the countryside matrix of fields, hedges and small copses, and those restricted to 'island' biotopes, such as fens, heaths, unimproved grassland and the larger woods, within the matrix. The widespread species, of course, also occur in most of the island biotopes within their ranges. With the exception of the large tortoiseshell (and the inclusion of this very rare species as a butterfly of the general countryside is arguable, it could be regarded as a woodland species), the butterflies of the matrix are common species.

This division into 'matrix' and 'island' butterflies is not always clear-cut

Table 1.1 Current status of British butterflies. In some cases, allocation to a category is arguable and the most doubtful of these are in parentheses. In particular, towards the edges of their ranges, butterflies are more likely to be restricted to particular favoured sites. Information from a variety of sources, but especially Heath *et al.* (1984) and Emmet and Heath (1989). The changes in range (over the last 40 years or so) do not include some very recent minor changes, e.g. the silver-spotted skipper has colonized new sites in the last few years, but nevertheless has contracted in range in recent decades. Similarly, the holly blue has been, in the short term, highly unstable, but over a longer period seems to fluctuate within a more or less stable range. A species designated as 'stable' may nevertheless have become much rarer within its range. Such a classification can never be entirely satisfactory and should be regarded as our 'best approximation'

	Change in range		
Status	*Stable*	*Expansion*	*Contraction*
(1) Widespread and (mostly) common species of the countryside.			
Small skipper		×	
Large skipper		×	
Brimstone	×		
Large white	×		
Small white	×		
Green-veined white	×		
Orange tip		×	
Small copper	×		
Common blue	×		
Holly blue	×		
Small tortoiseshell	×		
(Large tortoiseshell)			×
Peacock		×	
Comma		×	
(Speckled wood)		×	
Wall		×	
Hedge brown		×	
Meadow brown	×		
Small heath	×		
Ringlet		×	
(2) Restricted to a relatively small area of the country, but widespread species of the countryside within that area.			
Essex skipper		×	
Scotch argus		×	

Status	*Change in range*		
	Stable	*Expansion*	*Contraction*
(3) Rare butterflies, restricted to particular biotope 'islands' such as heath, unimproved grassland or woodland; not generally present in the wider countryside.			
Chequered skipper (in Scotland)	×		
Lulworth skipper	×		
Silver-spotted skipper			×
Dingy skipper	×		
Grizzled skipper	×		
Swallowtail	×		
Wood white	×		
Green hairstreak	×		
Brown hairstreak			×
Purple hairstreak	×		
White-letter hairstreak	×		
Black hairstreak	×		
Small blue			×
Silver-studded blue			×
Brown argus	×		
Northern brown argus	×		
Chalkhill blue	×		
Adonis blue			×
Duke of Burgundy			×
White admiral		×	
Purple emperor			×
Small pearl-bordered fritillary			×
Pearl-bordered fritillary			×
High brown fritillary			×
Dark green fritillary			×
Silver-washed fritillary			×
Marsh fritillary			×
Glanville fritillary	×		
Heath fritillary			×
Mountain ringlet	×		
Marbled white			×
Grayling	×		
Large heath	×		
(4) Migratory species			
Clouded yellow			
Red admiral			
Painted lady			

(5) Extinct species during this century
Chequered skipper (in England)
Black-veined white
Large blue (reintroduced)

and a species may occur in the general matrix in one part of the country and be more restricted in another. One such example is the orange tip, which breeds generally in the countryside in the south of Britain, but is more or less restricted to the vicinity of riverbanks in parts of the north (Courtney, 1980); another is the speckled wood, which is a butterfly of hedges and banks in the west, but occurs mainly in woods in many other parts of its range.

Restriction of the rarer butterflies to nature reserves or other isolated areas within the countryside is most marked in the south and east of England, where most land outside reserves is used intensively. Recently, areas of agricultural land have been taken out of cultivation and allowed to develop as rough grassland and scrub. At present, such land seems to provide habitats only for the commoner butterflies, but it is possible that some of the rarer species will benefit eventually.

The fact that a particular butterfly population occurs on a nature reserve is not a guarantee that its future is secure. A chance event, such as unusually adverse weather, may extinguish it and, if the site is remote from other populations, recolonization may be virtually impossible. In addition, particular types of management of reserves may be necessary to maintain suitable conditions for a rare butterfly species. These topics are discussed further in the final chapters of this book.

The decline of rare butterflies has been effectively documented by repeated distribution records and maps (Heath *et al.*, 1984). The loss of a rare butterfly from a site will often entail its loss from the 10 km square in which that site lies; its distribution map changes and the loss is clear. The widespread butterflies present a more difficult problem. For example, the meadow brown, one of the commonest butterflies, could disappear from thousands of sites and yet its apparent distribution, based on presence or absence in 10 km squares, would be largely unchanged. The same is more or less true of some 20 widespread species (Table 1.1), especially in the southern half of Britain.

It seems probable, some would say certain, that dramatic declines in the abundance of some of the common butterflies have occurred in recent decades. Such changes can be inferred from a knowledge of the biotopes in which the butterflies are found and the extent of destruction of these biotopes. The most obvious example is the loss of unimproved grassland (i.e. permanent grassland, largely untreated with herbicides and nitrogenous fertilizers). Even such species as the meadow brown and small heath, with larvae which feed on grasses, are usually scarce or absent on improved grassland (Thomas and Lewington, 1991), but are often common on unimproved grassland. Less surprisingly, the common blue and small copper, which are characteristic of unimproved grasslands, are inevitably absent from improved grassland or short-term leys where their main food plants (birdsfoot trefoil and sorrels, respectively) are absent. In spite of the fact that many of the common butterflies have undoubtedly become less abundant within their ranges, there have been some recent notable expansions of range. These expansions are discussed below.

1.4 SUCCESSFUL BUTTERFLIES AND EXPANSIONS OF RANGE

The recent history of British butterflies is not entirely a catalogue of extinctions, range contractions and declines in abundance. Some species appear to be flourishing and several have expanded their ranges in recent years. Not unexpectedly, most of the more successful butterflies are those which can breed within the farming landscape (Table 1.1), but there are also some of the rarer butterflies which are, at present, faring relatively well.

There are few agricultural areas which do not provide sufficient semi-natural vegetation to support a number of butterflies and a few species thrive in areas of intensive farming. The large and small white butterflies are pest species which feed almost entirely on cultivated brassicas; they are very common butterflies even though efficient chemicals are now available to control them. It is also likely that some of the nettle-feeding species, notably the small tortoiseshell and peacock, are more abundant in areas of intensive arable cultivation than elsewhere in the country (Chapter 5). Nettles benefit from soils rich in nitrates and phosphates; they flourish at the margins of arable fields where the soils are enriched by the fertilizers applied to crops.

A spectacular case of range expansion of a woodland butterfly is that of the white admiral. After an earlier contraction of range in the last century, this butterfly spread rapidly, during the 1930s and 1940s, from a small area in southern England to as far north as Lincolnshire; it has maintained or even extended this range in recent years. The likely causes of the spread of the white admiral are discussed in some detail in Chapter 12. The comma also spread dramatically in the middle years of this century, after an earlier decline in the nineteenth century. Like that of the white admiral, the advance of the comma appears to be continuing and it has colonized sites in the monitoring scheme. In the countryside of southern Britain, the comma is now a reasonably common butterfly, but in Sussex, for example, there were only about six recorded sightings between 1830 and 1930 (Pratt, 1981).

Perhaps the most surprising recent distributional changes have been the expansions of range of several species that have always been considered to be common in southern Britain. There is good evidence that the Essex skipper, small skipper, large skipper, orange tip, peacock, speckled wood, wall, hedge brown and ringlet have all had periods of expansion in the middle and latter part of this century. These are almost all of the widespread butterfies except those which already occupy virtually the whole of Britain. In some cases, the expansions have continued over several decades, in others they have only become obvious quite recently. This information comes from a variety of sources, but some of the most compelling is found in recent accounts of butterflies in counties of the north of England (Derbyshire (Harrison and Sterling, 1985); Yorkshire (Sutton and Beaumont, 1989); Northumberland and Durham (Dunn and Parrack, 1986)). For example, Sutton and Beaumont write of the small skipper in Yorkshire, 'Formerly confined to the east of the county, this species has undergone a range expansion westwards and northwards since the early 1970s. Now widespread and locally common, except possibly in the far west, it is often the most common butterfly in wild grassy areas'. In nearly all cases, these butterflies are recolonizing areas from which they disappeared in the nineteenth century or during the early years of this century.

As will be described later, these changes are not obviously associated with improving climatic conditions and an explanation presents a challenge to the ecologist. It is possible that, when the effects of weather on butterflies are more fully understood, the explanation will indeed lie there.

1.5 NORTHERN BUTTERFLIES

In the north and west of Britain, especially in Scotland, the recent loss of butterflies has not been as severe as in England, although there have been declines of some species. Thomson (1980) shows that the loss of species in Scotland has been greatest in the southeast where agriculture is relatively intensive and the loss of biotopes for butterflies has been similar to that in England.

There are several butterfly species that are restricted to, or mainly associated with, northern or other upland areas of Britain. These butterflies are the chequered skipper (which occurred in the English midlands until the 1970s; Farrell, 1975); northern brown argus, mountain ringlet, Scotch argus, and large heath. Most of these species are abundant in their particular habitats and their immediate future seems reasonably secure. Losses have been greatest, or at least most noticeable, amongst those, such as the large heath and northern brown argus, with outlying populations in the north of England. These outlying populations have often been lost because of drainage of wetlands or other improvement of marginal agricultural land. As in the south, some butterflies in northern Britain have expanded their ranges in recent years; the speckled wood, ringlet and orange tip have made notable advances and the Scotch argus has also spread considerably in some areas (M.R. Young, personal communication).

1.6 MIGRATORY BUTTERFLIES

In addition to the resident British butterflies, there are three intermittently common migrants to this country, the red admiral, painted lady and clouded yellow, and the monitoring scheme provides abundant data on the first two of these. There are also rarer migrants, such as the Camberwell beauty (*Nymphalis antiopa*), Queen of Spain fritillary (*Argynnis lathonia*), pale clouded yellow (*Colias hyale*) and Berger's clouded yellow (*Colias alfacariensis*), but, so far, there is little or no information from the monitoring scheme on any of these species.

These migratory butterflies, whether common or rare, are unable to overwinter in Britain, except in the case of the red admiral (Chapter 9). Individuals of the three common species fly to Britain virtually every year, from further south in Europe or from Africa. Once here, they may then breed through the summer. There is no indication that numbers of the painted lady and red admiral have declined during this century, but the clouded yellow has been rare in recent years, except for a remarkable immigration in 1983. In addition to these three species, which may be regarded as the major migrants, migration is thought to play a part in the lives of some other species, especially the large and small whites. The nature and frequency of the migrations of these two whites has been the subject of some research and debate, and evidence from the monitoring scheme is discussed in Chapter 9.

1.7 CONCLUSIONS

It is clear that the distribution of British butterflies is in a state of flux. The majority of our species has changed in status, in many cases radically, in the last hundred years. Are such rapid changes unusual, or are they typical of previous centuries? Dennis (1977) has argued convincingly that all of our butterflies colonized or recolonized Britain after the last ice-age. It is easy to assume that there was subsequently, particularly in historic times, a period of relative floral and faunal stability, after the main woodland clearance had been made and the mediaeval pattern of land use had been established. Almost certainly, there was no such stability; changes in the intensity of land use are likely to have continued to cause major changes in the abundance of many species of wildlife, including butterflies. It is even possible that land-use changes, such as the early loss of high forest and the extensive drainage of wetlands in the seventeenth and eighteenth centuries, caused extinctions of some butterflies even before those species were identified as British residents. Emmet and Heath (1989) provide a recent review of the evidence on whether species such as the scarce copper, *Lycaena virgaurea*, and the purple-edged copper, *Lycaena hyppothoe*, were once resident in Britain.

Whether or not historic changes in land use led to changes in the status of butterflies, climatic variations undoubtedly have caused the abundance and distribution of species to wax and wane through the centuries (Dennis, 1977). The importance of climate to butterflies can be seen by simple inspection of their ranges (Heath *et al.*, 1984) and has been emphasized recently by Turner *et al.* (1987). For example, over 40 species have been recorded in many 10 km squares in Hampshire and Dorset in the extreme south, while in Scotland, even in the lowlands, fewer than 20 species is typical. Other factors, in addition to temperature, may be partly responsible for the relatively poor butterfly fauna of the north, but there is little doubt that the role of temperature is dominant. We know that temperatures have fluctuated widely over recent centuries (e.g. Manley, 1974) and can be certain that butterfly distributions responded, even though we have no indication of the extent of the changes before fairly detailed, if largely anecdotal, information became available some 150 years ago.

Thus the various strands of evidence suggest that, while the extent of decline of butterflies in the last few decades is atypical, change rather than stability is the norm. The Butterfly Monitoring Scheme has been in existence for only 15 years and the Lepidoptera Recording (distribution) Scheme for a few years longer. In relation to changes in the status of butterflies over the centuries since the last ice-age, the recent periods of data collection may seem ludicrously short. Nevertheless, in the last two decades many, quite major, changes have occurred, in both the abundance and distribution of British butterflies. These decades are likely to be in part typical of the decades and centuries before, but also likely to show special features related to the ever increasing influence of humans on the environment. The difficulty lies in judging the relative importance of normal background factors and of abnormal factors, whether local or widespread, caused by humans.

Even during the period of the Butterfly Monitoring Scheme, concern about habitat loss and the decline of butterflies has increased and there have also been new worries, such as the likelihood of climatic warming. If we are entering a period of rapid climatic change, the insects may be amongst the first organisms to respond and the butterflies, because they are monitored, may be amongst the first in which a response is recognized.

Given the current declines of many rare butterflies and the expansions of range of several other species, together with the likelihood of climatic warming in the years ahead, there can be no doubting the importance and fascination of monitoring this group of insects.

2

Aims and methods of monitoring

2.1 AIMS

The original aim of the Butterfly Monitoring Scheme was, quite simply, to provide objective information on changes in the abundance of butterflies. Such information is important for conservation and contributes to an understanding of the population ecology of butterflies. The basic information relates to the individual sites which are monitored, but it is a further aim to integrate the results from individual sites to gain a wider, synoptic, picture of changes in abundance.

As the scheme has developed, it has become clear that a considerable range of information has been acquired on local distributions of butterflies, colonization of sites, extinction, migration and flight-periods. The analysis and interpretation of these data may now be added to the original aims.

2.2 REQUIREMENTS OF A MONITORING METHOD

A method used for monitoring should be quick, simple to use, and provide good estimates of population size for all of the species present in an area. Clearly, no such perfect method exists for butterflies or for any other group of animals, and a method has to be adopted that is less than ideal.

The most commonly used and most satisfactory methods of estimating the population size of insects are based on capture-mark-recapture techniques. Most such methods require a minimum of two sampling occasions, usually on different days. On the first occasion, insects are captured and marked in some way (usually with a small spot of paint, or coloured marker pen); on the second occasion recaptures are made to assess the proportion of marked individuals in the population, and so to estimate population size. If all goes well, a population estimate is obtained for one species over one time period. Frequently, however, problems arise. For example, it may be found that marking individuals affects their behaviour and so their chance of recapture, as has been found with some butterflies (Singer and Wedlake, 1981; Morton, 1984). Apart from such difficulties, capture-mark-recapture methods are too demanding in time and labour to be considered for monitoring on a national scale, although, particularly when there are many recapture dates, they can provide a range of information on birth rates, death rates and other population parameters (e.g. Begon, 1979).

For many insects, trapping in flight has proved a valuable monitoring method. National monitoring schemes in Britain are run for moths, using light traps (Taylor, 1979), and aphids, using suction traps (Taylor, 1977), together comprising the Rothamsted Insect Survey. Butterflies have been trapped in Malaise traps (a tent-like structure which permits the entry of flying insects, but not their exit) in a garden in Leicester (e.g. Owen, 1975), but these traps are most suited to small insects which occur at higher densities than butterflies. Similarly, Walker (1991) has used large net traps to intercept migrating butterflies, and, for this special purpose, they have proved very effective. However, for general monitoring of butterflies, all traps have the major disadvantage that the size of the catch is likely to depend greatly on behaviour, particularly whether individuals are dispersing or migrating.

The chosen method for monitoring must be a compromise between the ideal and what can be achieved in practice. The transect counts used in the Butterfly Monitoring Scheme are such a compromise. In choosing this method, rather than capture-mark-recapture, the main sacrifice is of information on absolute population size. The transect counts provide only an index of population size, which can be used to measure change in abundance over time, not an estimate of the number of individuals in a population.

2.3 THE MONITORING METHOD

2.3.1 Background

During the 1960s and early 1970s, N.W. Moore, then head of research into the effects of toxic chemicals on wildlife, made regular counts of butterflies along the southern edge of Monks Wood, the nature reserve adjoining the Monks Wood Experimental Station (Moore, 1975). His results were perhaps somewhat disappointing; because of the travel commitments of his job, the counts could not be made with sufficient frequency to provide reliable information on fluctuations in abundance. However, Moore's main aim was to develop a means of monitoring butterflies and the methods used in the Butterfly Monitoring Scheme are essentially a modification of his method. The only important differences are that, in monitoring scheme transects, the counts are more frequent, the routes walked are longer and more varied than that used by Moore, and the reliability of the counts have been fully tested.

In addition to the early work on butterfly transects by Moore, a variety of reasons and interests led four workers to begin trials of the monitoring method in Monks Wood (Pollard *et al.*, 1975). Transect counts of hoverflies (Diptera: Syrphidae) in Monks Wood had provided information on relative abundance from year to year. The method used for hoverflies differed from that for butterflies; individual flies were not identified in flight in the field (although this is possible for many hoverfly species), but were captured for later identification in the laboratory. Nevertheless, the consistency of results and the simplicity of the method augered well for similar transect studies of butterflies. D.O. Elias was then the warden of Monks Wood and was interested in improving the scientific basis of monitoring in the wood; M.J. Skelton worked in the Biological Records Centre and was involved with the distribution records of butterflies, while J.A. Thomas was studying the black hairstreak in the wood and had also experimented with transect counts of butterflies on flowers along one of the main rides.

The concept of transect counts is basically very simple and it is perhaps not surprising that other workers, unknown to both Moore and ourselves, had earlier used very similar methods. Goddard (1962), Douwes (1970) and Ekholm (1975) all used variations on the method adopted for the Butterfly Monitoring Scheme. Ekholm's work was most relevant to that of the scheme. He used transect counts in monitoring butterfly numbers in southern Finland for over 20 years, although the locations where he recorded changed over the period. He obtained some fascinating data on gross changes in abundance in a region near to the northern limit of some butterfly species.

2.3.2 Transect routes

The early trials of the monitoring scheme methods were in Monks Wood in 1973 (Pollard *et al.*, 1975). As has since become standard at other sites, a route through the wood was selected to include some of the better butterfly areas (Figure 2.1) and some that were not so good. The route included moderately shaded rides (tracks), sunny rides and one of the two large fields in the wood. Dense woodland compartments, which occupy a large part of the wood, but in which few if any butterflies are seen, were not included. In retrospect a sample of the woodland compartments should have been included to make the route more representative of the wood as a whole. But, at the time, most of the woodland compartments were heavily shaded and few, if any, butterflies were seen.

There would have been two advantages from including woodland areas: firstly, the data obtained would have confirmed that conditions were indeed unsuitable for butterflies and there would be no need to rely on assertion; secondly, conditions within woodland change over time, either naturally or

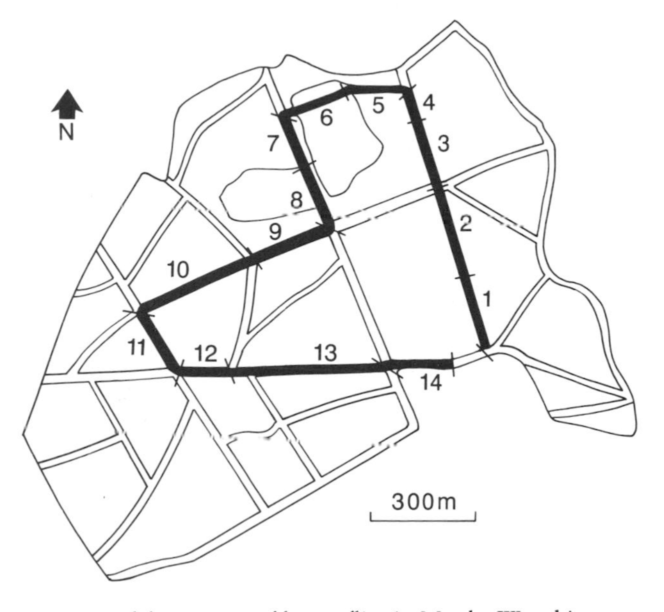

Figure 2.1 Route used for counts of butterflies in Monks Wood in Cambridgeshire, where monitoring began in 1973. The route, which is divided into 14 sections, runs mostly along the major rides (tracks) and through one of the two large fields (section 6).

through management, and in Monks Wood the opportunity to monitor such change has been missed. In the neighbouring conifer plantation of Bevill's Wood, one section through a compartment of young conifers was included. After a few years, the recorder had great difficulty fighting a way through the growing trees, but as compensation, there is a clear record of the gradual exclusion of butterflies as shading increased. This problem of representativeness, encountered right at the start of the monitoring programme, is one which will recur in this book.

The transect routes are divided into a maximum of 15 sections; at Monks Wood there are 14 (Figure 2.1). Separate counts are made for each section. The division into sections provides information on the local distributions of butterflies, although, for various reasons which are discussed later, interpretation of these distributions may be complex. The choice of sections is partly determined by features of the route, for example a major ride intersection is an obvious division, and partly chosen to reflect local changes in vegetation type. Where there are long, uniform, stretches of vegetation along part of a transect route, these stretches are usually divided arbitrarily into sections, as the uniformity may eventually be lost by management or by some unexpected event, such as fire.

The total length of the Monks Wood route is about 3 km and it takes between 60 and 90 min to record. Experience suggests that this length is about the optimum, although there is considerable variation in the lengths of the transect routes at the sites in the monitoring scheme.

The precise width of the sections is decided by the recorder, but once adopted is not changed. It is recommended that the width should normally be no more than 5 m. The adoption of a common width for all routes would in many ways have been an advantage, but presents practical problems. For example, in a clearly-defined woodland ride 6 m wide, bounded by banks or ditches, it is very much simpler to use the whole width of the ride for recording, using the banks or ditches as distinctive edges, rather than to judge whether or not a butterfly enters a narrower recording band within the ride. Indeed, as butterflies in woodland rides are often concentrated in a narrow belt of vegetation at the ride edges, adoption of a specific width could itself lead to anomalies. For example, if only one of the edges is included in the recording band in a wide ride, but both edges in a narrower ride, a comparison of numbers of butterflies in the two rides might be quite misleading.

This variability in width of routes at some sites makes strict comparison of the abundance of butterflies in different sections invalid. Often, however, differences between sections in numbers of butterflies are so great that differences in section widths are of little importance. In the one detailed study of local distributions in which the widths of woodland rides was included as a variable (Greatorex-Davis *et al.*, 1992), variation in width was found to be unimportant.

2.3.3 Recording

All butterflies seen within the bounds of the route, and within an estimated distance of 5 m ahead of the recorder, are counted. Various rules of recording are used to standardize the results obtained by different recorders as much as possible (Hall, 1981). No attempt is made to count butterflies flying high above the recorder, so that species which habitually fly in the canopy, such as some of the hairstreaks, are recorded only on the rare occasions when they are at ground level. Some recorders, especially in open countryside, find it helpful to imagine themselves in a moving 5 m box and to record all butterflies that they see within the box.

The aim of recording is not to count all butterflies present while the recorder traverses the route; rather, those butterflies are counted which the recorder sees while walking at a steady pace. Thus, no special effort is made to see any butterflies which may be settled out of direct sight in dense vegetation. Sometimes a single conspicuous butterfly will fly along in front of the recorder, in and out of the recording area. If the recorder is sure that this is indeed the same individual, it is counted only once; if the recorder is unsure, it is counted again.

The recording season comprises the 26 weeks from 1 April to 29 September inclusive. At least one walk should be completed in each recording week. Almost inevitably, some weeks are missed because of bad weather, especially early and late in the recording year; inevitably also, occasional weeks are missed because of illness, holidays (if a substitute recorder is not available) or other commitments. Nevertheless, a complete record of 26 (or more) counts is the aim and has been achieved by many recorders. In most years, a few butterflies fly in October and in some years, at some sites, the number may be quite large. This information is lost to the scheme. The recording season is restricted mainly because of the time required for processing data and writing reports, ready for the beginning of the next recording session. At the beginning of each season, an account of the features of greatest interest in the results from the previous year is sent to the recorders; in this way an interest in the accumulating results is encouraged. In general, such encouragement does not seem to be necessary, as many recorders take a keen and critical interest in the scheme and its results.

As there are only 58 species of butterfly in Britain, identification is generally quite easy. A person unfamiliar with butterflies can cope with monitoring quite readily, especially if he, or she, starts recording at the beginning of a season. Many recorders have found it an excellent and gradual way to learn their butterflies. Early in April there are only two or three easily identified butterflies, and new species appear gradually from week to week. At Monks Wood there have been only two difficult identification problems: the separation of the small and Essex skippers and of the small and green-veined whites. These skippers are not readily identifiable unless netted; no attempt is made to separate them and the data are combined under 'small skipper'. The two white butterflies are difficult only if in rapid flight. It is recommended that recorders carry a net and, if a 'problem' individual can be easily caught, the recorder may stop to identify it; counting also stops until the walk is resumed. Otherwise, if the recorder is unsure of an identification, the butterfly is recorded as the commoner (at the time) of the possible options.

At other sites there are different identification problems. For example, where pearl-bordered and small pearl-bordered fritillaries, or high brown and dark green fritillaries, or female common blues, chalkhill blues and Adonis blues are present together at a site, similar problems occur. In practice, differences in flight periods, behaviour, state of wear (age) of individuals, preferred areas for flight and other clues can often be used to decide between possibilities. As with the white butterflies and skippers, extra caution is needed in assessing results for these pairs or groups of species.

The males and females of some butterfly species are conspicuously different in appearance; extreme examples are the orange tip and the 'blues' with blue males and brown females, but it is also easy to distinguish, while recording, male and female hedge browns, meadow browns, and several other species. In the monitoring scheme, the sexes are not recorded separately, but some recorders do so and add to the interest of their recording and to the information gained.

Most butterflies can, on occasion, be seen in flight from early morning until almost dusk, especially when temperatures are high. For example, one of us (E.P.) recently saw a holly blue in flight at 7.30 a.m. on an April morning, with a white frost on the ground, while, frequently, meadow browns are seen soon after dawn. However, some species seem to be restricted in the time of day during which they will fly, irrespective of temperature, and so the counts are further standardized by restricting recording to around the middle of the day. The period specified is between 10.15 and 15.45 British Summer Time (09.15 and 14.45 Greenwich Mean Time).

Butterfly flight is to a large extent dependent on temperature; in very cool weather they do not normally fly, irrespective of time of day. The requirements for recording are that counts may be made if the shade temperature is over 17.0°C, irrespective of sunshine; between 13.0 and 17.0°C, counts may be made if at least 60% of the walk is made in sunshine (i.e. 60% of the sections are, when walked, predominantly sunny). The lower temperature limit is reduced to 11.0°C at northern upland sites, where recorders tell us that butterflies are more tolerant of low temperatures.

The temperature criteria for recording are a compromise. If the limits are set too high, a recorder would often have difficulty in finding suitable conditions for a count. Even using the chosen criteria, it is sometimes impossible to find a suitable day for a count early and late in the recording year. If the limits are set too low, it is likely that counts would be made when only a small proportion of the butterflies present were flying.

Wind speed during a count is estimated, using the Beaufort scale, and noted on the recording form. It is generally inadvisable to record in wind speeds in excess of force 5 ('small trees in leaf begin to sway'), but, as the effect of wind varies greatly in different biotopes, no rigid standard is set. In woodlands, for example, butterflies are able to find sheltered areas for flight even in the strongest winds, while a stiff breeze may prevent flight on exposed cliffs and downs.

Finally, information on management and other events thought likely to influence butterfly populations should be recorded for future reference. In some cases, annual photographs of the transect route have been taken and, at a very few sites, data on changes in vegetation have been collected.

2.3.4 Results: index of abundance

The recording method provides counts for the complete route and for each section separately. The recording form illustrated in Figure 2.2 is typical in that it shows that the highest numbers of several species occur in a few favoured sections. At the height of the season, recording in such sections requires great concentration. Some recorders memorize the counts while walking and note the totals at the end of each section, or part of a section, but others use tally marks to score each sighting.

Over the flight period of a particular butterfly species, a series of counts is obtained; those for the ringlet at Monks Wood in 1973 are illustrated in Figure 2.3. In this case there were several counts in each recording week. The sequence of counts, first increasing then declining, results from the appearance of individuals in the population, either by emergence of local butterflies from pupae or by immigration, and subsequent disappearance, by death or by emigration. The ringlet, the example used here (Figure 2.3), is believed to have fairly discrete local populations (Thomas and Lewington, 1991) and, if this is indeed so, the rise and fall of counts will reflect the emergence and death of individuals.

The counts are used to calculate an index of abundance. The index is the sum of the mean weekly counts (Figure 2.3). If, as at most sites, just one count each week is made, the index is then the total number of butterflies seen. In the case of the ringlet, as with many other butterflies (Appendix A), there is one generation of butterflies each year. In other species, such as the wall and green-veined white, there are usually two generations with distinct flight periods and an index of abundance is calculated for each. In warm summers, these and most other 'bivoltine' butterflies may produce another generation; this third generation is usually small and often overlaps with the second and we include it with the second generation index. In a smaller number of cases, such as the speckled wood and small heath, the flight periods of different generations overlap so much that no separate index values can be calculated. For such species we sum all the weekly counts to give a single annual index. The brimstone and peacock butterflies are also special cases, in that the same butterflies have flight-periods before and after hibernation but they have only one generation a year; separate autumn and spring index values are calculated. Other species which overwinter as adults, the comma and small tortoiseshell, are given a single annual index because the sequence of generations may be complex and variable.

Missing counts for one or more weeks present a problem in the calculation of the index of abundance. A subjective judgement is made as to whether there are sufficient counts to permit the calculation of an index; if there are judged to be sufficient, missing counts are estimated as the mean of the preceding and succeeding counts. This subjective element is undesirable,

BUTTERFLY CENSUS

Year	73	Date	9.7	Recorder	E. P.
1-2		3-5		6-8	
		Site name	MONKS WOOD		
9-11		12-17		18-19	
Start time	11.00	End temp °C	19°	% Sun	90	End wind speed	2
20-23		24-26		27-28		29	

	30-32	33-35	36	39	42	45	48	51	54	57	60	63	66	69	72	75	78-80
Section		1	2	3	4	5	6	7	8	9	10	11	12	13	14	15	Total
Brimstone	54																
Common blue	106																
Green-veined white	99		1			1		1	1					2			6
Hedge brown	76					1	7	8	1		2						19
Large skipper	88	1							1								2
Large white	98																
Meadow brown	75	1	1	4		1	8	12	13	4	10	1		2			57
Orange tip	4																
Peacock	84																
Red admiral	122																
Ringlet	8	1	1	1		3	7	17	17		38			2	1		88
Small copper	68																
Small heath	29						3										3
Small skipper	120						4		1								5
Small tortoiseshell	2																
Small white	100						1										1
Wall	94																
White admiral	64		1											1			2
Marbled white	78						1										1
W I hairstreak	113		2														2
Section		1	2	3	4	5	6	7	8	9	10	11	12	13	14	15	
Sunshine		S	S	S	S	S	S	S	C	S	S	S	S	S	S		

Notes: HEDGE BROWNS ALL ♂♂ PLEASE TOTAL EACH SQUARE

Figure 2.2 Completed recording form for Monks Wood on 9 July 1973, starting at 11.00 a.m. Recorder E. Pollard. The small printed numbers, including a code for each butterfly species, are aids to data processing. The marbled white was then very rare in the wood and soon became extinct.

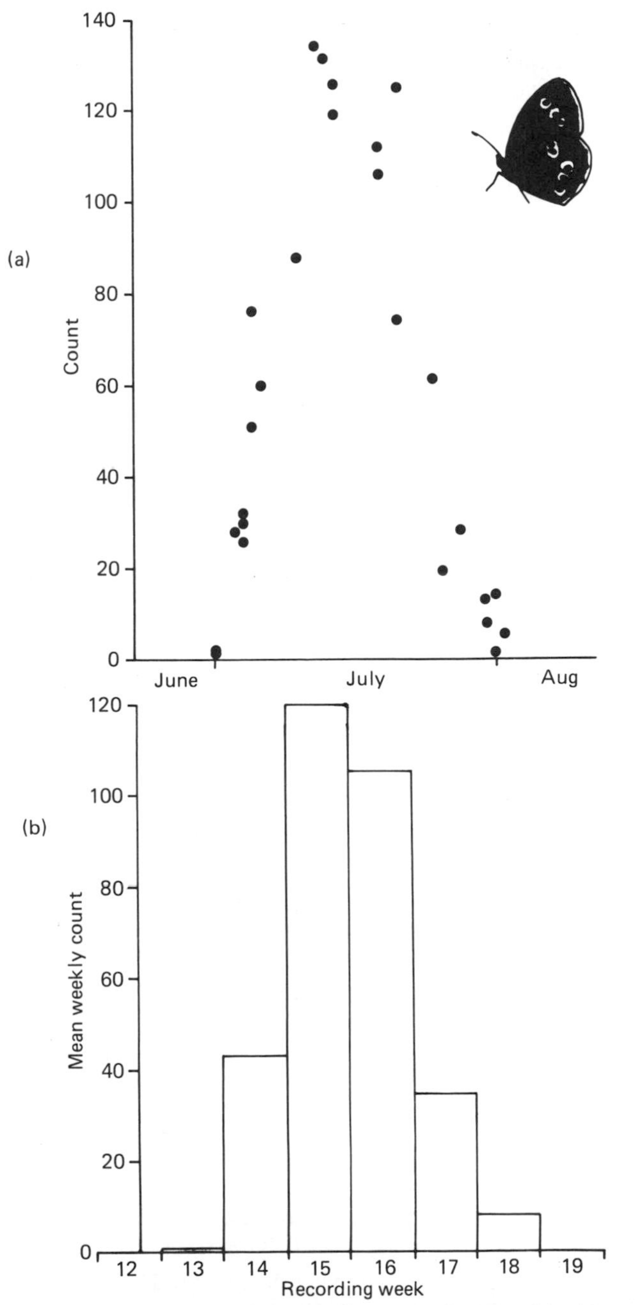

Figure 2.3 (a) Individual counts of the ringlet in Monks Wood in 1973. (b) Mean count in each recording week. The index of abundance for the generation is the sum of the mean weekly counts and in this year was 307.

as judgements can change over time; indeed although the decisions have been made by one individual over the whole recording period (E.P.), there is no doubt that standards have become stricter and more data rejected, as the scheme has progressed.

The index of abundance is used to assess change in abundance from generation to generation. The validity of the index depends on the assumption that a count for a particular species is a more or less constant, although unknown, proportion of the number of butterflies present on the route. For each species, the proportion seen is likely to be different; some butterflies, such as marbled whites, are conspicuous, others, such as dingy skippers, are much less likely to be seen. Thus the index values cannot easily be used to compare the abundance of different species.

A sequence of counts, such as those in Figure 2.3, follows a curve that can be fitted mathematically. If a particular model for the curve is assumed, an index of abundance, together with a measure of its variability, can be calculated from the available data and so the problem of missing counts avoided. Such a method has not been adopted in the monitoring scheme, partly because it could not be applied to species with complex, overlapping, flight-periods, or when numbers are very low, but there is potential for its use with some abundant species.

Provided there is no net migration and a particular emergence pattern is assumed, it is possible to use a sequence of counts to calculate the death rate (and so longevity) of adult butterflies, in addition to an index of abundance (Zonneveld, 1991). Zonneveld's method suggests death rates very similar to those calculated using more laborious capture-mark-recapture methods, but his method entails the assumption that death rates are constant over the flight-period.

The index of abundance used in the monitoring scheme is simple to calculate and requires very few assumptions about the structure of the data. However, the index is strictly an index of butterfly flight-days i.e. a combination of the number of butterflies and the length of their lives. Variability in weather may perhaps influence length of life from year to year (Chapter 10) and so affect the relationship between the index and population size, but any such effects are likely to be small.

2.4 CONCLUSIONS

In this chapter, the recording method used in the Butterfly Monitoring Scheme has been outlined and the use of the counts to provide an index, a measure of relative abundance, has been described. This index is reliable if its variability, due to effects of weather or other factors, is small compared with changes in the index caused by changes in population size. The validation of the index is considered in Chapter 3.

– 3

Validation of the monitoring method

3.1 INTRODUCTION

The early years of recording at Monks Wood, from 1973 to 1975, were, in part, devoted to establishing whether the transect counts of butterflies provided a robust method for monitoring. In particular, it was necessary to know to what extent counts varied with different recorders, and to be sure that variation was not so great, at different times of day and in different weather conditions, that the counts reflected these factors rather than differences in numbers of butterflies.

Validation of the counting method took various forms. For a few species, population estimates were made in the same area and over the same period as the monitoring counts. It was also possible, as is described later, to assess differences between recorders, effects of temperature, sun and windspeed directly from counts. The frequent counts at Monks Wood were ideal for such analyses. Apart from the need to establish, without doubt, that the monitoring method gives information on population changes, analyses may reveal unexpected features about the ecology of butterflies and so be of more general interest.

3.2 COMPARISON OF TRANSECT COUNTS WITH ABSOLUTE POPULATION ESTIMATES

Transect counts should be closely correlated with estimates of population size made at the time of the counts. Capture-mark-recapture studies were performed to estimate the population size of three butterfly species at Monks Wood in 1974 and 1975, for comparison with transect counts.

The choice of species for marking studies is limited. Some butterflies fly very strongly and are difficult to catch; others are migratory or range over wide areas, in which cases populations cannot easily be defined or estimated. The three selected Monks Wood species, the small heath, ringlet and green-veined white (Pollard, 1977), were all reasonably abundant, quite easy to capture and were believed to occur in sedentary populations.

The method used to estimate population size was based on the frequency of capture of individuals during one day (Craig, 1953; Eberhardt, 1969). Sampling was along a predetermined route through the sampling area, capturing butterflies and marking each captured individual with a small spot of rapidly drying paint. The circular route was completed many times during the day. On each subsequent capture of a marked butterfly, a further small mark was given, so that the number of captures of each individual was known; thus a frequency distribution of the number of individuals captured once, twice, three times, etc., was obtained. It is likely, for several reasons, that some individuals will be more susceptible to capture than others and in such cases the frequency distribution of captures is expected to follow a particular (truncated geometric) distribution (Eberhardt, 1969). The zero class of the frequency distribution, i.e. the number of butterflies not caught at all, is estimated and the total population can then itself be estimated. The advantage of using this method is that a population estimate is obtained from a single day's sampling.

Population estimates were made for the small heath on nine occasions in 1974 and 1975 in East Field, Monks Wood, the field through which the transect runs (Figure 2.1). The small heath is a butterfly mainly of open grassland and in Monks Wood the majority of records have been from this field. The population estimates for the small heath were therefore compared with transect counts made in the field. Population estimates of the ringlet were made on five occasions, also in East Field, in 1975. The ringlet is generally more common in woodland rides than in the field and the population estimates were compared with counts in both the wood and the field. The ringlet, unlike the small heath, has a single discrete flight-period each year and the results are presented to show the passage of a generation of butterflies (Figure 3.1).

For both the small heath and ringlet, the correlation between population estimates and transect counts, in the week of the estimates, was reasonably close (Pollard, 1977).* One difficulty in interpreting the results is the considerable variability of the population estimates. This is a frequent problem in capture-mark-recapture studies, partly because sample sizes are usually small, and it is rarely possible to estimate insect population size with precision.

For the third species, the green-veined white, estimation of population size, using this method, proved impossible. The captures were made around a circuit of woodland rides, as the green-veined white is relatively scarce in the fields. Unexpectedly, as more individuals were captured and marked during a day, the proportion of marked individuals that were recaptured failed to increase; thus the estimates of population size continued to

* Correlation coefficients: small heath, $r = 0.55$, 34 counts; ringlet, $r = 0.76$, 27 counts; both highly significant, $P < 0.001$.

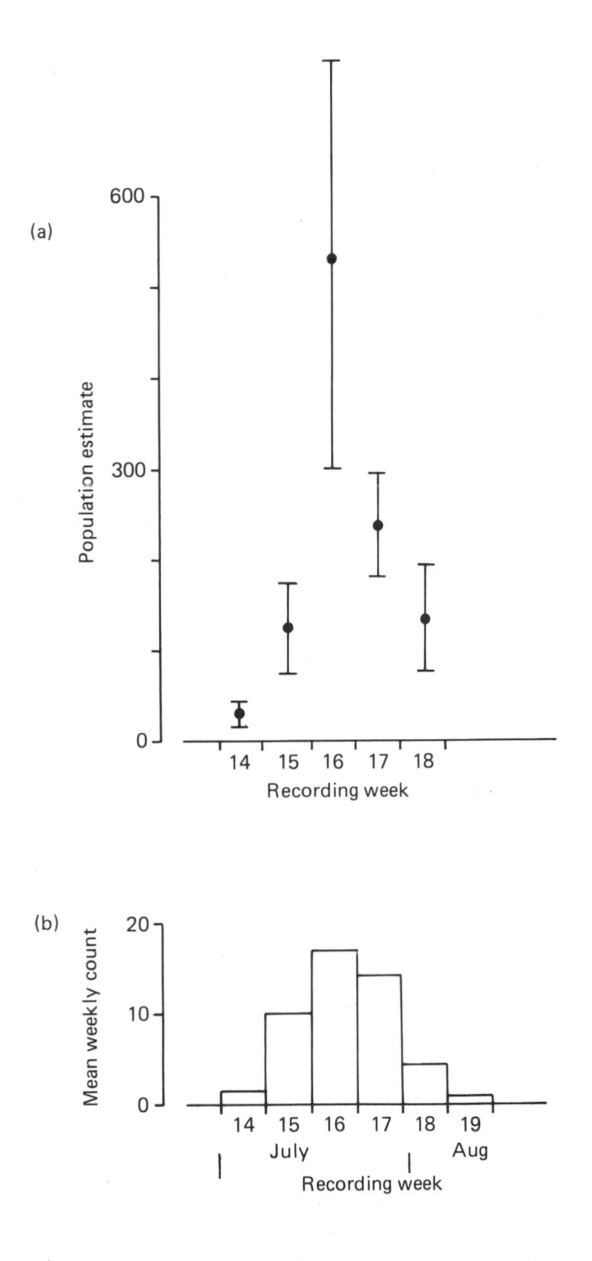

Figure 3.1 (a) Population estimates ± SE (each based on frequency of capture on a single day) of the ringlet in East Field, Monks Wood in 1975. (b) Mean transect count in the field in the weeks of the population estimates. Correlation between population estimates and the 27 individual transect counts in the weeks of the estimates: $r = 0.76$, $P < 0.001$. Figure redrawn from Pollard (1979b).

increase. It was thought at the time (Pollard, 1977) that the butterflies were moving through the wood, i.e. marked butterflies were leaving the wood and new, unmarked, ones were flying in. It was concluded, therefore, that the population was less sedentary than had been assumed before starting the study. Subsequently, Warren (1981) obtained similar results for a woodland population of the wood white, which was known to be resident in his study area. Warren's interpretation was that marked butterflies tended to move out of the circuit of rides in which captures were made into other nearby rides and so became unavailable for capture. These marked, but unavailable, butterflies remained in the larger population of the wood, perhaps only a few metres from the sampling route, while other unmarked butterflies replaced them in the sampled rides. In retrospect, it seems that a similar situation occurred in the study of the green-veined white in Monks Wood. The population was, after all, likely to have been sedentary. This method of estimation of population size requires that the chance of capture of any individual butterfly should remain the same during the period of sampling. In the study of the green-veined white, this condition was not met and we had allowed ourselves to be misled by one of the many problems that bedevil capture-mark-recapture studies.

In addition to these studies in Monks Wood, other workers have shown close correlations between population estimates, by capture-mark-recapture methods, and transect counts. Species which have been studied are the marsh fritillary, Adonis blue, Glanville fritillary, small blue and silver-spotted skipper (Thomas, 1983a), the heath fritillary (Warren, 1987c) and the wood white (Warren *et al.*, 1986). In Thomas's studies and that of the heath fritillary, the wider aim was to compare absolute population sizes at a range of different sites in the same season. To achieve this, the transect count method was modified in two ways: (1) the transect counts were 'calibrated' against population estimates, using the capture-mark-recapture method, at a few sites for each species; and (2) at each site the transect route zig-zagged across the whole area known to be occupied by the population. In this way, counts per unit length of each route could be converted into approximate population estimates. It is the calibrations which are relevant in the context of this chapter, as they provide the same type of comparison between transect counts and population estimates as in the earlier Monks Wood studies.

As in Monks Wood, Thomas (1983a) compared transect counts with population estimates made by Craig's (1953) method of frequency of capture in a single day. The agreement between transect counts and population estimates for all these species (e.g. Figure 3.2) was very close (closer than in the Monks Wood studies), and further helped to confirm the reliability of the transect counts. Warren's (1987c) study of the heath fritillary included similar comparisons of transect counts and population estimates by Craig's (1953) method. Again, the agreement was close.*

In the wood white study (Warren *et al.*, 1986) the population estimates were by the method of Jolly (1965). Using this method, butterflies at a single site are captured, and each individual given a mark by which it can be identified, on a series of days over the duration of the flight-period; the presence of individuals from previous days is recorded on each sampling day. Various population parameters, including population estimates through the flight-period, can then be calculated. Close agreement was found between transect counts, made on the sampling days, and these population estimates. Males and females were recorded separately and agreement was equally good for both sexes.†

In the transect counts of the wood white in Warren's study, more than twice as many males as females were recorded, although sexing of pupae and the capture-mark-recapture estimates both suggested that the true sex ratio was close to 1:1. As in most butterflies, wood white males spend more time in flight than do the females, are seen more often by recorders and have

* Correlation coefficient: $r = 0.90$, nine sampling occasions (several sites; $P < 0.001$).
† Males, $r = 0.83$, $n = 21$, $P < 0.001$; females, $r = 0.85$, $n = 20$, $P < 0.001$; data for 2 years combined.

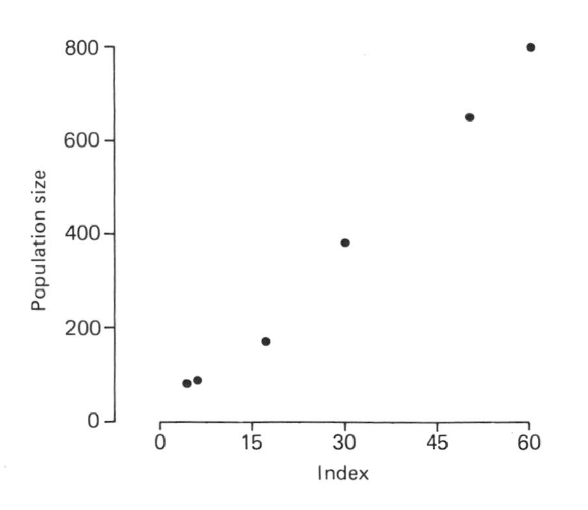

Figure 3.2 Population estimates of the Glanville fritillary plotted against an index of population size. The population estimates were each based on frequency of capture on one day and the index based on transect counts (corrected for the area occupied by the population and the length of the transect route) made on the same day as the estimates (Thomas, 1983a). The close relationship between the index values and population estimates is evidence that variation in counts is largely determined by variation in population size. Figure redrawn from Thomas (1983a).

higher counts. This difference between the sexes may be assumed to be consistent from generation to generation and population changes, shown by transect counts, should be unaffected.

3.3 RECORDER DIFFERENCES AND THE EFFECTS OF TIME OF DAY AND WEATHER ON COUNTS

Simple inspection of the counts can provide convincing evidence of the reliability of both counts and index values. For example, the index of abundance of the ringlet in Monks Wood has fluctuated greatly, with a lowest value of 8 in 1977 and a highest of 769 in 1987. The individual counts in those 2 years (Table 3.1) suggest that it is impossible for the large and consistent differences in counts to be caused by variation in weather conditions during the counts, or indeed by any factors other than very large changes in population size. Differences in abundance between years as pronounced as this are unusual, but in general the consistency of counts within a year, and frequent large differences between years, provide similar evidence that changes in the index values reflect population changes. Although such inspection of results gives confidence that the monitoring method is robust, there are nevertheless likely to be detectable, if minor, effects of weather and other factors on the counts. Some examples of such effects are described in this section.

In Monks Wood, in 1973 and 1974, three recorders, E.P., D.O. Elias and M.J. Skelton, made frequent counts throughout the season. These data can be used to test for differences in recording between these individuals and also for effects of weather and time of day. As an example, the counts for the ringlet at Monks Wood in 1973 (Table 3.1) and 1974 were analysed. Each of 57 counts was classified according to the following factors: year; week in the flight-period (only the central 6 weeks considered); time of day (the 5 hours of recording); wind speed (five categories); sunshine (two categories, mainly sunny or mainly cloudy); recorder (three individuals); and temperature.

Of these factors, year and week of recording made the only substantial contribution to the variability in counts (Table 3.2, full analysis unpublished) and the 'recorder effect' was negligible. Time of day and temperature had small, but significant, effects. The contribution of temperature to the overall variability of counts was negligible, but time of day had a larger effect, with more ringlets recorded in the afternoons. However, the effect of all other factors was small, relative to that of the week and year of recording.

In all, 91% of the variability was accounted for by the factors included in the analysis, with the other 9% due to other unknown factors. These unknown factors might include chance variation in the number of butter-

Table 3.1 Counts of the ringlet in Monks Wood in 1977 and 1984. The two years contrast sharply in the size of counts, but inspection of the data shows that temperature and sunshine, the factors most likely to influence the proportion of butterflies present that are actually seen and counted, had little effect. For example, the highest count in 1984 was 195, in cool weather with no sunshine; the highest count in 1977 was 6, at the same temperature with the sun shining. There can be no doubt that the differences in counts reflect real differences in the size of ringlet populations in the 2 years. (From Pollard *et al.*, 1986)

1977			*1984*		
Count	*Temperature (°C)*	*Sun (%)*	*Count*	*Temperature (°C)*	*Sun (%)*
1	20.5	30	1	18.5	100
1	19.0	90	2	18.5	90
1	18.0	40	12	22.5	70
1	17.5	20	74	22.5	80
1	20.0	0	123	18.5	70
1	16.0	60	195	18.0	0
2	17.5	0	178	18.0	0
5	17.0	100	93	21.0	90
6	18.0	100	92	24.0	0
3	25.0	20	86	23.0	100
2	21.5	30	43	21.5	100
2	21.0	100	65	21.0	50
1	18.0	0	33	19.0	50
1	19.0	70	13	20.0	100
			26	18.5	70
			11	20.0	90
			17	23.5	0
			14	22.5	100
			6	24.0	40
			14	28.0	100
			2	24.0	100

flies flying across the route, variation in the numbers actually seen by the recorder (e.g. temporary loss of concentration!), counting errors, effects of weather conditions other than those measured, and so on. This analysis suggests that, for the ringlet at Monks Wood in these years, the main cause of variation in counts was the number of butterflies present at the time, with other factors of minor importance. The lack of any apparent difference between recorders was encouraging and perhaps rather surprising. However, these particular recorders had worked closely together on the development of the recording method and would be expected to make similar counts.

Table 3.2 Influence of various factors on counts of butterflies. Results from analysis of variance of counts for four species at sites in the monitoring scheme. The influence of each factor is indicated by r^2 (percentage of total variation in counts that is accounted for by each factor†). In general, only the week and year of recording make substantial contributions to the variability of counts; these factors reflect the changing abundance of butterflies through each flight-period and the fluctuations in abundance from year to year. Significant effects: indicated by $^*P < 0.05$, $^{**}P < 0.01$, $^{***}P < 0.001$.

	Percentage of total variation accounted for by:						
Species and site	*Recorder*	*Time of day*	*Temp.*	*Wind speed*	*Sun*	*Week*	*Year*
Ringlet, 57 counts, Monks Wood, 1973–74	0	4*	0	2	0	70***	15***
Large skipper, 155 counts, Monks Wood, 1973–86	–	1	1	1	0	17***	54***
Common blue, 67 counts, Tentsmuir Point, 1978–90	–	2	0	7***	0	5	68***
Scotch argus, 71 counts, Loch Garten, 1978–90	–	1	0	4	0	43***	23

† r^2 values for factors included in order of decreasing significance

Similar analyses have been conducted with only a few other sets of data, selected haphazardly from the vast number available. One species selected was the large skipper, because it was considered in an earlier study (Pollard *et al.*, 1975) to fly mainly in sunshine. Analysis of 155 counts of the large skipper at Monks Wood by just one recorder (E.P.), from 1973 to 1986, failed to demonstrate an effect of sun (Table 3.2). As with the ringlet, the year and week of recording had highly significant effects, but, of the other factors, only the effect of temperature was approaching significance and its contribution to the overall variation in counts was negligible.

At Tentsmuir Point in Scotland (Chapter 13) strong winds were found to depress counts of the common blue (Table 3.2). The effect of wind speed on counts was, in this case, larger than that of the recording week (the common blue here has a single generation, with a long flight-period and with no strong peak of emergence), but much less than the difference between years. Temperature, sunshine and time of day did not have significant effects. Tentsmuir Point (see Site Studies, Chapter 13) is an exposed coastal site where, if anywhere, wind may be expected to have a substantial impact on counts; this analysis suggests that the effect of wind was to increase the variability of the index values from year to year, but not to such an extent as to obscure the underlying population changes. There is no doubt that the flight of the common blue (and of other species on exposed sites) is reduced in strong winds.

Counts of another northern butterfly, the Scotch argus, at a more sheltered inland site, Loch Garten, showed no significant effect of wind speed (Table 3.2). Once again, the major part of the variability was related to the week and year of recording. The contribution of the year of recording to the variability of these counts was rather small, probably reflecting the relatively small population fluctuations of the Scotch argus at this site.

These analyses of the effects of conditions during recording show, as expected, that a range of factors affect counts, but that their influence is likely to be small relative to that of differences in the size of populations. In each example, over 70% of the total variation is accounted for by the variables measured and the large majority of this is by the week and year of recording i.e. by the number of butterflies available to be counted at the time. This is a basic requirement of a monitoring method and there is no doubt that it is sufficiently precise to provide the required information.

There is enormous scope for further similar analyses of the data held in the scheme, for example to examine the effect of temperature on butterfly flight. However, the value of such analyses may be reduced because of the difficulty in separating effects on the actual number of butterflies in flight from effects on the number of sightings; the two are not necessarily the same. For example, strong winds may depress butterfly activity, but they may also reduce the ability of a recorder to see butterflies against a background of moving vegetation.

Although the Monks Wood data on the ringlet, analysed above, did not show differences between recorders, experience suggests that such differences do sometimes occur. In interpreting data from individual sites, a change of recorders is one of the factors taken into account. There is no training of recorders; each person learns the method from an instruction booklet (Hall, 1981). Although these instructions are written to be as unambiguous as possible, there is undoubtedly an element of individual interpretation. At a few sites the recorders are different each year. Provided that local supervision of these recorders is thorough, and consistent from year to year, such a system seems to produce satisfactory results; it is, however, less satisfactory than recording by a single individual over many years.

3.4 SUNSHINE, TEMPERATURE AND BUTTERFLY FLIGHT

Sunshine appears to have little impact on the size of counts in the small number of species that we have examined in detail (Table 3.2). However, in many accounts of the natural history of butterflies, statements are made to the effect that some butterflies fly only in sunshine. For example, in a recent excellent and authoritative book on British butterflies (Emmet and Heath, 1989) it is stated or strongly implied that eight species require sun for flight. The species include the common blue (which showed an effect of sunshine in data from Tentsmuir Point; Table 3.2), the chalkhill blue and speckled wood. These three species occur at many monitored sites and abundant data are available from counts in sun and shade.

If there was such an absolute requirement for sunshine for butterfly flight, our guidelines for recording would be inappropriate (recording is permitted in warm conditions, irrespective of sunshine). Fortunately, only a few further examples (Table 3.3) are needed to show that the three listed species are seen in cloudy conditions, provided the temperature is sufficiently high.

It is likely that all butterflies are capable of flight in shade, as the importance of sunshine is to raise the temperature of an individual above a threshold necessary for flight (e.g. Pivnick and McNeil, 1987). If the temperature in the shade is sufficiently high to enable a butterfly to reach this threshold, then flight will not require sunshine. Nevertheless, in very poor light, flight may be suppressed (Douwes, 1970); after all, butterflies very rarely fly at night, whatever the temperature.

Strictly, the monitoring counts are of butterflies either in flight, or settled on vegetation and flowers; thus the counts in sun and shade, in Table 3.3, may not relate to butterflies that are all in flight. However, virtually all settled butterflies seen on counts are only temporarily at rest between

Table 3.3 Counts of butterflies in sunny and shady conditions. Some authors (see text) consider that these species fly only in sunshine; these few examples of counts in sun and shade, separated by only a few days, show that butterflies are recorded in the absence of sun

Species and site	*Date*	*Count*	*Sun (%)*	*Temperature (°C)*
Speckled wood,	25 June 1990	14	0	22.0
Monks Wood	2 July 1990	8	70	17.0
Common blue,	1 Sept. 1990	22	0	22.0
Dyfi	3 Sept. 1990	20	80	19.0
Chalkhill blue,	22 Aug. 1990	13	0	23.0
Pewsey Down	25 Aug. 1990	6	100	24.0

periods of flight, and can be regarded as capable of flight.

It is well known that butterflies vary their posture, when settled, to increase or decrease their exposure to the sun. The grayling, for example, basks with the undersides of its wings at right angles to the sun when it needs warmth, and parallel to the sun's rays to prevent over-heating (Findlay *et al.*, 1983). Such behaviour suggests that there is an optimal temperature for activity. Indeed, in very hot weather, butterflies may seek shade. For example, while collecting bales of hay from a field in the afternoon of a very hot day in August 1989 (temperature 34°C), one of us (E.P.) noticed that few, if any, meadow browns flew in the open field, but many were sheltering in the small patches of shade cast by individual bales. Such high temperatures occur only rarely in this country and are probably of little consequence for monitoring. However, upper temperature limits for transect counts would be required in hotter parts of the world.

3.5 CONCLUSIONS

The particular method of transect recording was developed because data on many species could be collected over a large area in a relatively short period of time. The method was originally thought to be inferior to capture-mark-recapture methods, which are not possible for monitoring because they require so much time and labour. However, the results from monitoring suggest that the transect-count method may be more robust than capture-mark-recapture techniques if the main need is to estimate change in population size rather than absolute population size. The monitoring method may be preferable because (1) transect recording requires fewer assumptions about the behaviour of individual butterflies and about the structure of populations; and (2) there is no handling of butterflies, with possible subsequent effects on their behaviour. In addition, transect recording can be used when populations are very large, with thousands or tens of thousands of individuals, or for butterflies which are very difficult to capture; in such cases it is virtually impossible to mark enough individuals to ensure an adequate number of recaptures.

The transect recording method has been used as an alternative to mark and recapture adult butterflies in life-table studies of the white admiral (Pollard, 1979a) and the wood white (Warren *et al.*, 1986); in both cases the information on numbers of adults was well related to counts of other stages of the life cycle.

In spite of the evidence presented in this chapter, that the index values are not greatly influenced by weather conditions on the day of recording, there is little doubt that the minimum conditions for recording are close to the flight thresholds of some species. Counts made just above these limits are likely to be low counts. Presumably, the apparent lack of impact of weather on counts and index values is partly because most counts are made in conditions well above the minima, and partly because large fluctuations in population size overwhelm other influences.

The studies described in this chapter relate mostly to individual counts, rather than the index values, which are the sums of series of counts. Unless there are major differences in the longevity or movement of butterflies in different years, validation of the transect counts can be equated with validation of the index values. As the index values are based on several or many counts, they should be more robust estimators of relative abundance than are individual counts.

The early Monks Wood trials of the monitoring method in 1973 were sufficiently promising for further pilot trials at several nature reserves in eastern England in 1974 and 1975. The main aim of these trials was to discover whether reserve wardens could accommodate butterfly counts in their working week, without undue disruption of normal work. One might expect recording to be a minor addition to the host of other tasks undertaken by the wardens, as counts usually take only 1–2 hours per week. Indeed, Goldsmith (1991) comments that butterfly monitoring is a classic example of good monitoring on a small budget, because it requires of the recorder 'little more than the dedication of a weekly lunch hour'. However, the recorders know well that butterfly monitoring is a major commitment; the 1 or 2 hours in a week required for recording are not predictable. In a generally cool and cloudy week, the sun may shine and the temperature rise only for a brief period, when other important matters are at hand. In spite of such difficulties, all the wardens in the pilot trials coped successfully with recording and in 1976 the national scheme was begun.

– 4

Sites, site selection and 'national' monitoring

4.1 INTRODUCTION

In the first year of the Butterfly Monitoring Scheme in 1976, counts were made at 34 sites. These included the sites in eastern England where pilot trials were conducted in 1974 and 1975, 24 other nature reserves, two farmland areas, two conifer plantations and a disused railway cutting. Over the following years the number of sites increased steadily until 1979, then remained at 75–80 through the 1980s until 1989, when there was a further small increase to the present 90–95 sites.

Many of the early recorders were wardens of National Nature Reserves. These wardens obviously have a special interest in the sites they manage, are ideally placed to undertake regular recording and are excellent naturalists; this combination of qualities makes them the ideal recorders. As discussed in Chapter 1, several British butterfly species have a high proportion of their populations on National Nature Reserves or other wildlife areas; therefore if the scheme is to include some of these rarer British butterflies, many nature reserves must be included. The predominance of nature reserves in the monitoring scheme has continued (Table 4.1) and the addition of some sites in 1989, specifically because of the rare species they contained, further added to the number of reserves.

A high proportion of monitored sites is in the south of England (Figure 4.1). This concentration in the south and southeast partly reflects the distribution of potential recorders and partly the distribution of the butterfly species themselves. Another concentration of sites is in the east of England, stemming from the origin of the scheme at Monks Wood in Cambridgeshire. Further north, the number of butterfly species declines and butterfly recording also becomes more difficult. These factors are of course connected; conditions which are difficult for recording are difficult for butterfly flight and, presumably, have some influence on the survival of species in a region. Although there are fewer sites in northern areas, these few are particularly important in that they provide data on population

Table 4.1 Ownership of sites used for calculation of 'all-sites' index values (Chapter 6). All sites recorded in two or more years. The sites are dominated by nature reserves of various types, with relatively little farmland and commercial forestry

	No. of sites
Nature Conservancy Councils	51
Local Wildlife Trusts, Woodland Trust	17
Royal Society for the Protection of Birds	13
Forestry Commission	7
National Trust and NT for Scotland	4
Local Authorities	2
Private	12

fluctuations under harsh climatic conditions and include some species not present in the south.

In this chapter, the range of sites in the scheme is described briefly, together with the representation of butterflies. The use of the sites for assessing national changes in butterfly numbers, and limitations in this use imposed by the particular selection of sites, are discussed.

4.2 SITES IN THE SCHEME

In spite of the preponderance of sites in southern England, those that have contributed to the scheme range from the Lizard Peninsula in Cornwall, to the islands of Skomer and Skokholm off the west coast of Wales, Lindisfarne off the coast of Northumberland, Inverpolly in the Western Highlands and Sands of Forvie in the northeast of Scotland. Murlough, an area of coastal dunes and heath, is the only site in Northern Ireland, although an independent scheme now operates there.

The biotopes present at the sites in the scheme (Table 4.2) include a broad selection of the deciduous woodlands, grasslands, wetlands and heathlands which make up the majority of our nature reserves. Similarly reflecting the distribution of nature reserves in Britain is the large number of coastal sites with cliffs, dunes and coastal heaths.

Perhaps the most important single biotope for butterflies in Britain is calcareous grassland, especially that on the chalk of southern England. Some butterflies, such as the chalkhill blue, Adonis blues, and the silver-spotted skipper are entirely or largely restricted to such sites. Many other rare and local species also occur on them. Chalk grasslands at Kingley Vale, Old Winchester Hill, Aston Rowant, Gomm Valley, Castle Hill, Wye, Lullington Heath, Swanage (Ballard Down), Pewsey Down and Martin Down have been recorded for all or most of the years of the monitoring scheme.

Figure 4.1 Location of sites in the Butterfly Monitoring Scheme. The sites are those currently in the scheme, plus all others which have been recorded for at least 4 years.

A further range of sites of special importance for butterflies are ancient deciduous woodlands and these too are well represented. These sites include Castor Hanglands, near Peterborough, and Wyre Forest in the west midlands. Several, such as Ham Street Woods in Kent and West Dean Woods in Sussex, and to a more limited extent Monks Wood, have areas managed as coppice to enhance their suitability for butterflies. Many ancient woodlands are now planted with conifers, and some such areas,

Table 4.2 Classification of sites according to biotope. In a few cases, where there are substantial areas of two biotopes at a site, or the categories are not mutually exclusive, sites have been included under two categories. The dominance of nature reserves is reflected by the large number of sites with semi-natural biotopes, rather than economic crops

Deciduous woodland	28
Conifer plantations	7
Chalk and limestone grassland	18
Heath	5
Wetland	12
Upland	6
Coastal cliffs, dunes and marshes	16
Farmland	8
Other	15

including Picket Wood in Wiltshire, Potton Wood in Bedfordshire and Bevills Wood in Cambridgeshire, have been monitored for many years.

The wetland sites in the monitoring scheme include such famous East Anglian reserves as Chippenham Fen and Woodwalton Fen (where the reintroduced large copper butterfly occurs), the National Trust's Wicken Fen, and Bure Marshes in the Norfolk Broads (Figure 4.2). Bure has a population of the swallowtail, a rare butterfly which occurs in Britain as an indigenous subspecies.

Several reserves belonging to the Royal Society for the Protection of Birds (RSPB), which are of course primarily selected for their importance for birds, have butterfly transects. One of these, Radipole Lake, a wetland area in Weymouth in Dorset, has been recorded since 1977, as has Northward Hill in Kent, where woodlands adjoin the High Halstow marshes of the Thames Estuary. In Wales, the RSPB reserve of Ynis-Hir (Figure 4.3), on the estuary of the River Dyfi, has an exceptionally rich butterfly fauna. In Scotland the famous 'osprey' reserve of Loch Garten also has a butterfly transect and the Scotch argus is abundant there.

Two sites with substantial areas of heathland are Walberswick on the Suffolk coast and Studland Heath in Dorset. At Studland the silver-studded blue, a rare butterfly largely restricted to heathland, is present in reasonable numbers; at Walberswick it has not been recorded on counts since 1984 but is still present nearby.

In the Scottish Highlands, suitable recording conditions may be so scarce in some summers that a complete monitoring record is impossible, and the same is true at Upper Teesdale in the Pennines. The highest altitude where counts have been made is at 800 m in the Eastern Highlands of Scotland on Ben Lawers, a mountain famous for its arctic-alpine plants and the only site in the scheme where the mountain ringlet occurs.

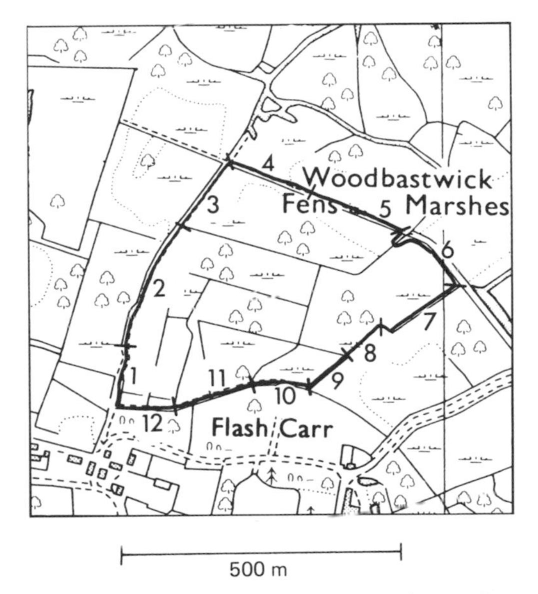

Figure 4.2 Butterfly monitoring transect at Bure Marshes in the Norfolk Broads. The swallowtail is restricted to the Broads and has been recorded regularly on this transect. In recent years, the white admiral has colonized the area.

There are few Welsh sites, but two, Oxwich on the Gower Peninsula, and Dyfi, another coastal area on the Dyfi estuary, have complete records since 1976, and Newborough Warren, in Anglesey, has been recorded since 1978. Oxwich in particular has a notable butterfly fauna, as its range of biotopes extends from coastal dunes and freshwater marsh to woodlands on limestone cliffs. An upland nature reserve in Wales, Craig y Cilau near Brecon, also has many years of counts, but has not been recorded since 1988.

The few farmland sites in the monitoring scheme include Woodwalton Farm, lying between the nature reserves of Monks Wood and Woodwalton Fen. Woodwalton Farm is a typical area of Cambridgeshire arable farmland; it has been recorded by Monks Wood recorders since the scheme began in 1976. Another arable farmland site is at Alresford in Hampshire, near to the famous Alresford watercress beds, while Batch Farm on the edge of the Cotswolds, Brockwells Farm in Gwent and Springhill Farm in the Weald of Kent are grassland areas which are not farmed intensively. The downland site at Swanage, although a designated Site of Special Scientific Interest, is commercially managed farmland.

There are a few sites in the monitoring scheme which cannot be easily fitted into the broad categories described above. For example, Hampstead Heath in north London has been recorded since 1978 and St Osyth in Essex,

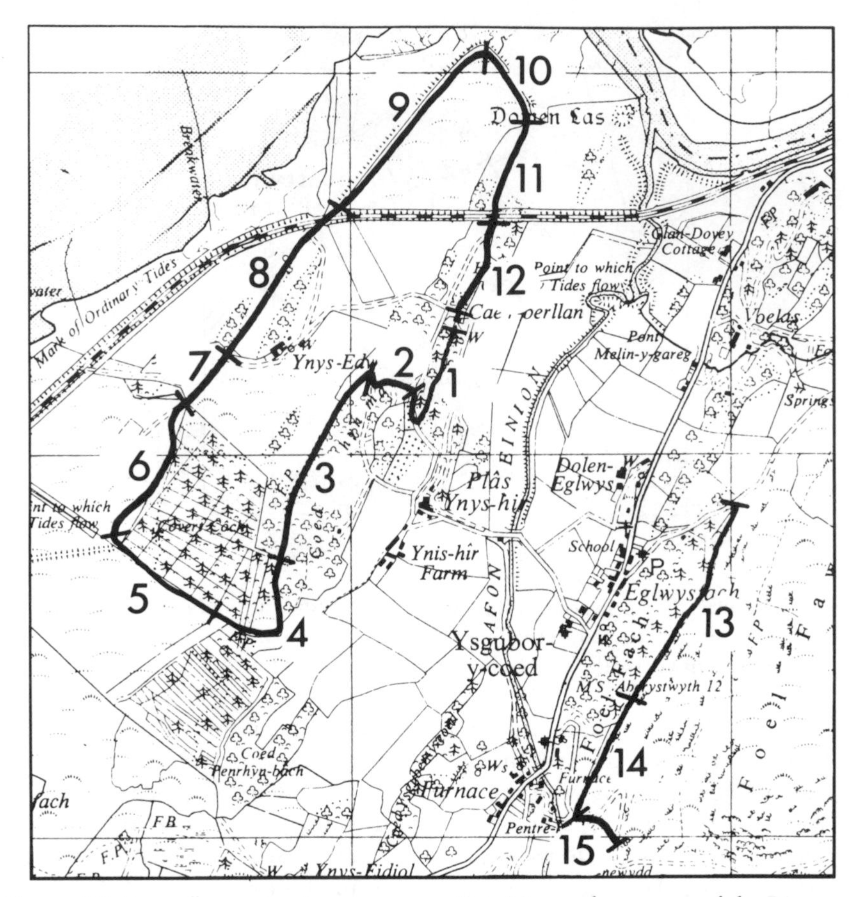

Figure 4.3 Butterfly monitoring transect at Ynis-Hir on the estuary of the River Dyfi in Wales. Although a 'bird' reserve, there is a rich butterfly fauna, including the silver-washed fritillary and green hairstreak, reflecting the wide range of woodland and open biotopes. Scale indicated by the 1 km square grid.

a revegetated landfill site (former waste disposal area) since 1983. Some land-use categories, which might be included, are missing from the scheme; for example, there is no recording for the scheme in suburban gardens, although these often contain breeding populations of some butterfly species and attract many others to visit flowers.

This brief survey has not included many sites which have provided long series of data for the monitoring scheme. All the sites have contributed to the results that are presented later in the book, and some of those omitted here are discussed in some detail in later chapters. All current sites and those that have contributed data for four or more years are listed in Appendix B.

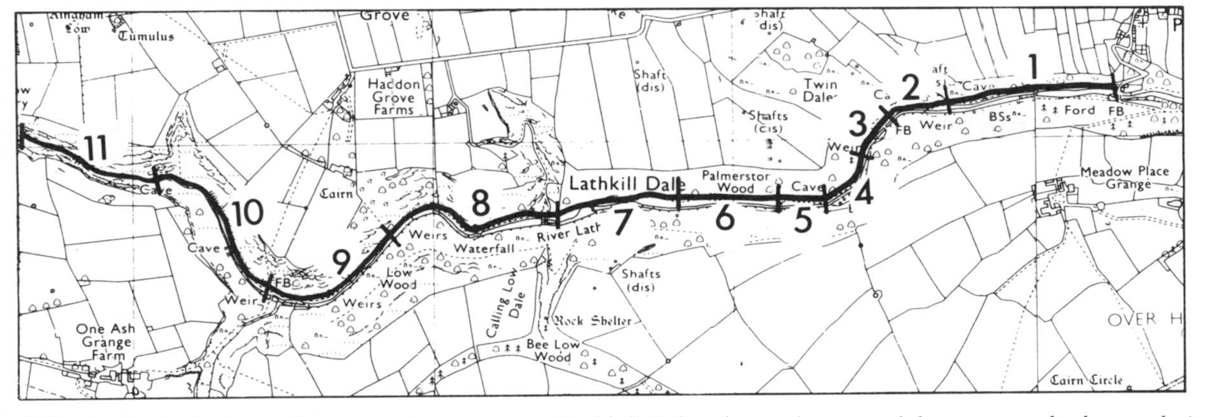

Figure 4.4 Derbyshire Dales butterfly monitoring transect at Lathkill Dale. The site has one of the most southerly populations of the northern brown argus. Large numbers of orange tips are recorded and the dale was colonized during the northward extension of the hedge brown late in the 1980s. Scale indicated by the 1 km square grid.

4.3 REPRESENTATION OF BUTTERFLIES

The butterflies divide clearly into two groups (Figure 4.5). In the example illustrated, one group consists of 21 species which were recorded at over 40 sites and the other of 32 species, which were recorded at 30 sites or fewer, with many of these recorded at between one and 10 sites. The details differ slightly from year to year, but the general pattern is similar to that shown.

In Chapter 1, the British butterflies were broadly classified, according to whether they occurred widely in the ordinary agricultural countryside or were restricted to 'island' biotopes within the matrix of the countryside. The grouping of species in Figure 4.5 is a reflection of this division. It is not surprising to find that the widespread resident butterflies, plus the common migrants, the red admiral and painted lady, are present at many sites in the monitoring scheme. It is, however, surprising that the separation of this group is so clear, especially as the sites in the monitoring scheme are strongly biased in favour of conserved areas and contain many more rare

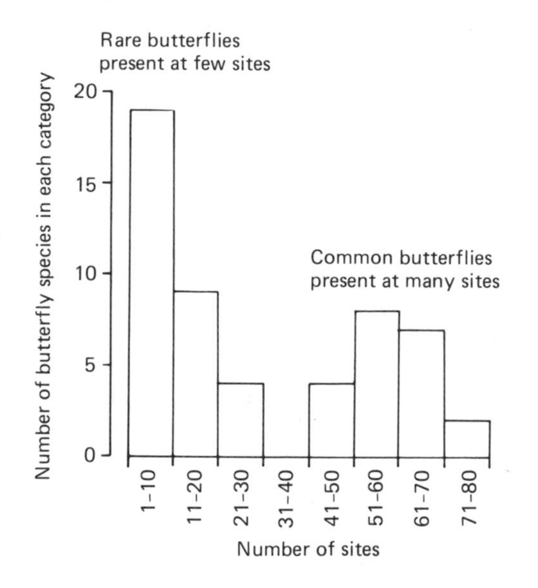

Figure 4.5 Histogram of the number of butterfly species in eight classes, based on the number of sites at which each species was recorded in 1990. The butterflies divide into two groups, one of rare butterflies present at few sites and the other of common butterflies present at the majority of sites. The division reflects the fact that much of Britain is farmed or otherwise intensively used, with small 'island' areas of richer vegetation which are often relics of former land use. The common butterflies occur more or less throughout, while the rare butterflies are, in general, confined to the 'islands'.

species than would a random selection of sites. In such a random selection, representative of the British countryside, the distinction between the common, widespread, butterflies and the rarer, localized, species would still be present, although a very large random sample of sites would be required for the rare species to be represented fully.

4.4 SITE AND ROUTE SELECTION AND NATIONAL MONITORING

The sites in the monitoring scheme are independent areas at which butterflies are monitored. However, an important aspect of the scheme is the integration of results from individual sites to provide some sort of overall picture of changes in numbers of British butterflies. In later chapters the integration (collation) of site index values, based either on all of the sites in the scheme or on the sites in a particular region, is described. In the context of such an overall picture, the composition of the sites and selection of transect routes are important. In this section, we discuss the potential, and limitations, of the use of data collated to provide a more general picture of changes in butterfly numbers.

4.4.1 Biases in site representation in the scheme

The Butterfly Monitoring Scheme depends on voluntary recorders who, in general, wish to record at sites rich in butterflies. If the scheme is to provide information on more species than the 20 or so common and widespread butterflies, then this concentration of effort on sites rich in butterflies is essential. However, a consequence is that the sites are not representative of the wider countryside in terms of the types of biotope that they include. As many butterfly species are restricted to southern Britain, the sites also tend to be concentrated in the south.

4.4.2 Effects of monitoring on monitored sites

On the farmland transect in Cambridgeshire, recorded by Monks Wood scientists, a green lane on the route initially provided a pleasant walk on a grassy track between hedges. Over a period of years the lane gradually became overgrown with scrub, and butterfly numbers declined. Eventually, it was discovered that normal practice on the farm was to mow the lane fairly frequently to prevent scrub encroachment, but since we had begun recording, the farm manager had left it uncut to prevent 'disturbance' of the butterflies. The lack of management, counter to the intention, was deleterious to the butterflies, but the point of the anecdote is to show that the monitoring of a site can influence its butterflies.

Such indirect effects of monitoring can be reduced if it is recognized that the possibility of such effects exists. However, only an entirely different approach to monitoring would eliminate them entirely. One possible alternative to the use of fixed routes is the selection of new sites each year, in effect taking a series of independent samples from the 'national butterfly population'. However, such an approach would present too many logistical problems for a scheme based on voluntary recorders and with few resources.

4.4.3 Use of collated index values from all sites

The effects of the predominance of nature reserves, and of other biases in the selection and subsequent management of sites, on the results of the monitoring scheme are unknown. There is no reason to suppose that the broad pattern of year-to-year fluctuations is affected, but it is probable that long-term trends of many species are different at sites in the scheme as compared, for example, with typical farmland. It would be surprising if most butterflies did not fare better on areas acquired and managed for wildlife than in the wider countryside, although there are many instances in which rare butterflies have been lost from nature reserves.

The biases should not be disregarded, but the results obtained by collation of data from the sites in the scheme are nevertheless of considerable interest, for the following reasons:

1. The short-term, year-to-year, changes shown by the collated data are likely to be the most suitable data for detecting effects of weather on butterflies.
2. For a group of highly mobile butterflies, such as the brimstone, peacock, small tortoiseshell, large and small whites, and the migratory red admiral and painted lady, the individuals seen on a nature reserve will almost certainly have developed as larvae outside the reserve, sometimes many kilometres away. The numbers seen on a reserve are likely to depend to a large extent on abundance in the wider countryside.
3. For some species, fluctuations from year to year and longer-term trends in numbers may be so large as to overwhelm local effects of site management. This may be the case in the future if climatic warming causes major changes in the distribution of butterfly species.
4. As several of the rarer butterflies and others, such as the dingy skipper and marbled white, have a substantial proportion of their populations on nature reserves or other protected areas, a monitoring scheme with a predominance of such sites is appropriate for these species. If butterfly species are shown to be in decline on sites which include many protected areas, this is clearly cause for concern.

– 5

Local distribution of butterflies

The favourite haunts of the marbled white are rough uncultivated grassy hill-sides, meadows and sometimes rough openings and outskirts of woods. It is usually very locally distributed, is often confined to a certain field in abundance and hardly found elsewhere in the neighbourhood.

F.W. Frohawk, *The Complete Book of British Butterflies*, 1934.

5.1 INTRODUCTION

The transect route, at each site in the monitoring scheme, is divided into sections and separate counts are made in each section. Thus, after the very first count at a site, the recorder has some information on where on the route particular butterflies occur, i.e. their local (or spatial) distribution.

In this chapter, results of monitoring at the downland site of Castle Hill are used to show some local distribution patterns of butterflies and the interpretation of these patterns. This is one aspect of monitoring which does not require regular recording year after year. Counts at a site made in one year can be repeated a decade or more later and the local distributions of butterflies compared. However, regular recording makes it possible to show whether distribution patterns remain essentially stable from year to year, change erratically, or show gradual shifts over the years. The nature of any changes may provide clues to the processes involved.

5.2 AN EXAMPLE: LOCAL DISTRIBUTIONS AT CASTLE HILL

The Castle Hill site is an area of chalk downland on the South Downs near Brighton. It is an open downland site with a transect route of a uniform 5 m. The route (Figure 5.1) includes sections through floristically rich downland, improved grassland and scrub. Butterfly monitoring at this site over a decade has been described by Pollard and Leverton (1991).

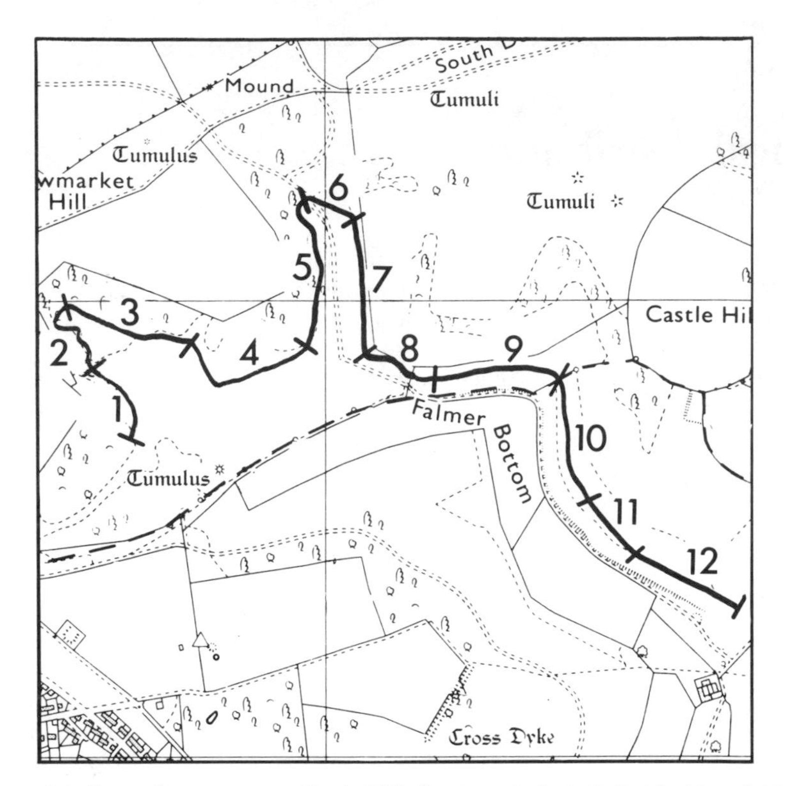

Figure 5.1 Butterfly transect at Castle Hill. Sections 1, 3, 4, 5, 8, 10, 11 and 12 are through herb-rich downland; section 2 through quite dense scrub; sections 6 and 9 through improved grassland, and section 7 along a track adjoining old downland outside the reserve and improved grassland within the reserve. Apart from very limited areas of rich downland, the reserve is largely surrounded by intensively farmed land. Scale indicated by the 1 km square grid.

5.2.1 Distributions

At Castle Hill on 25 August 1981 (Figure 5.2) the sun shone throughout the walk, which started at midday, and there was a force 3 wind. The count was fairly late in the season, but the chalkhill blue was at its peak and the recorder (R. Leverton) was lucky enough to see a clouded yellow in a year in which only 16 were seen at all sites. The first of a small late-summer generation of the Adonis blue was recorded in section 12. Such a 'snapshot' of the butterflies, in a particular area at a particular time, is amongst the basic information provided by the monitoring scheme.

On a given day, the recorded distribution may be influenced by factors such as wind speed and direction, intermittent sunshine during recording, and time of day. A more general picture at Castle Hill was given by the

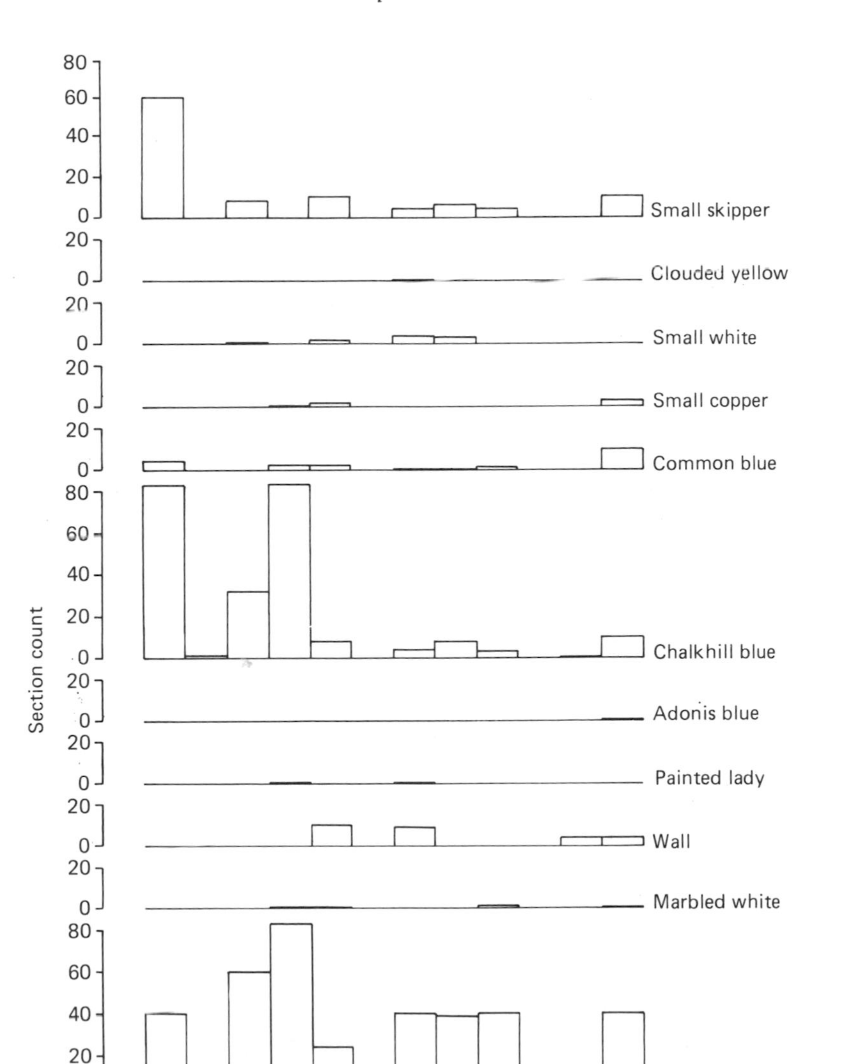

Figure 5.2 Distribution of counts of butterflies around the Castle Hill transect route (Figure 5.1) on 25 August 1981. The section counts are not corrected for the lengths of the sections.

distributions over the whole of the 1981 recording season (Figure 5.3). It is clear that, for the species present in moderate or large numbers on 25

August (Figure 5.2), the distributions on that day were similar to those over the whole season. The single-day snapshot, in this case, seems to have given a fairly reliable picture of the normal distribution of adult butterflies at the site.

Although the distribution of individual species varied considerably, in some favoured sections of the route, large numbers of several species were recorded regularly. Three of the four richest sections are old, unimproved, downland, while the other, section 7, is along a track close to similar rich downland outside the reserve. In some cases, species with very different habitat requirements, including those that prefer different sward heights, were abundant in the same sections. For example, both the Adonis blue and marbled white, which in Britain require very short and long turf, respectively (e.g. BUTT, 1986), were abundant in the downland sections 4 and 12. This apparent paradox probably occurs because the swards are not uniform in these areas, but are mosaics of short turf between tussocks of tor grass, and so both species thrive. Unfortunately, such a mosaic is difficult to maintain. At Castle Hill the spread of tor grass on the downland is a problem. If continued, it could eventually suppress the low-growing downland flora and associated insects; management, by grazing with cattle, is undertaken to attempt to prevent this from happening.

Two sections of the transect, sections 6 and 9, are through improved grassland which was formerly arable. Hence there were particularly low numbers of several butterflies, including the dingy skipper and chalkhill blue. All butterflies, except the large skipper and speckled wood were poorly represented in the scrub-dominated section 2.

5.2.2 Interpretation

It would be convenient for both reserve managers and butterfly ecologists if the distribution of adult butterflies gave a simple guide to the areas in which they breed. However, interpretation of patterns of local distribution depends on a knowledge of the mobility of butterflies (Appendix A). If a species is known to be very sedentary, the spatial distribution is more likely to give an indication of breeding areas; for example, the dingy skipper is very largely restricted to areas of old downland at Castle Hill, in spite of the fact that its main food plant in Britain, birdsfoot trefoil, occurs on both the unimproved downland and the grassland on recently abandoned arable land. This pattern of distribution has remained more or less stable for many years (Figure 5.4). As the dingy skipper is considered a sedentary butterfly it is probable that it breeds only on the downland, although even in this case such a restriction is not absolutely certain. Dingy skippers are not conspicuous, especially the females when laying eggs, and there may have been some unobserved oviposition in the improved grassland. Nevertheless, it is certain

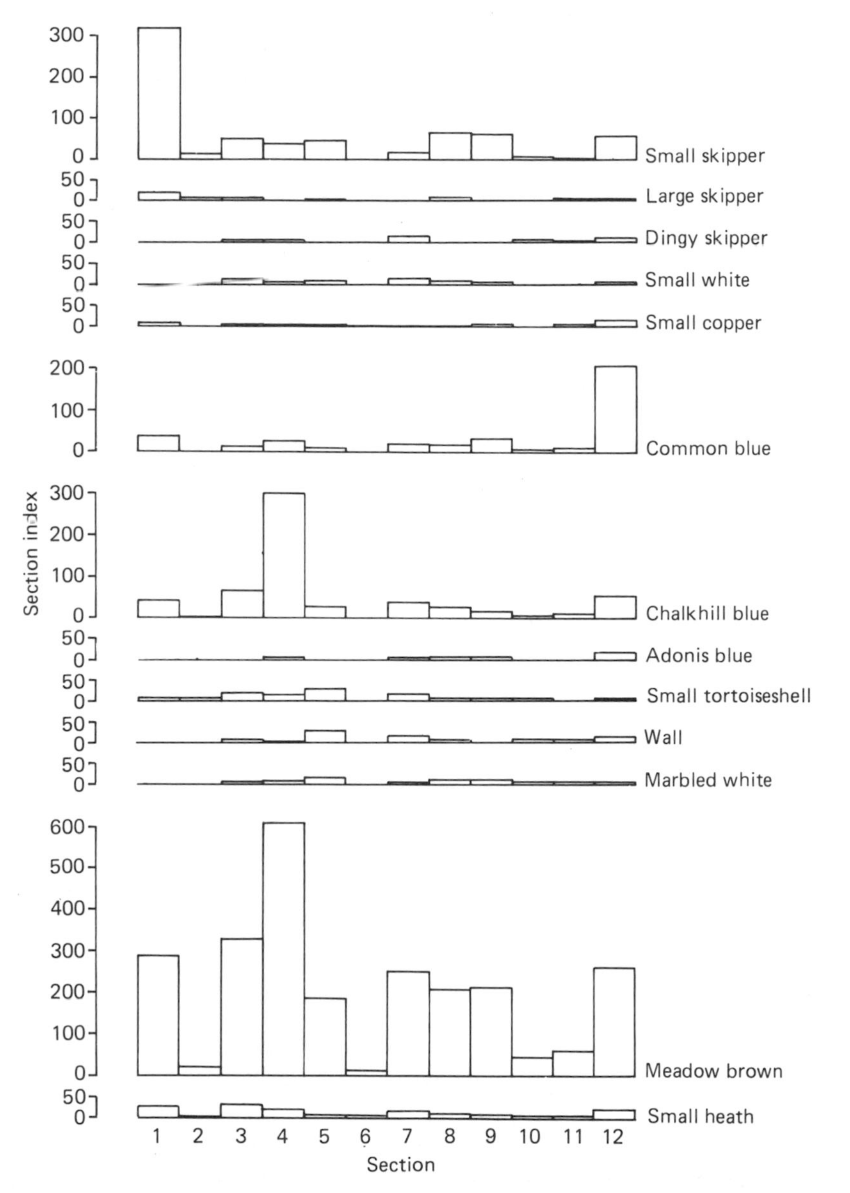

Figure 5.3 Distribution of section index values of butterflies around the Castle Hill transect route (Figure 5.1) over the 1981 season. The section index values are not corrected for section lengths. In addition to the species shown, the clouded yellow, large white, green-veined white, brown argus, red admiral, painted lady, peacock, comma, speckled wood and hedge brown were recorded in low numbers.

that the populations depend primarily on the ancient downland. Thomas and Lewington (1991) note that the dingy skipper prefers horseshoe vetch as its main food plant on downland, rather than birdsfoot trefoil. If this is the case at Castle Hill, the observed distribution would be explained, as horseshoe vetch is virtually absent from the improved grassland.

In contrast, the small white is known to be a very wide-ranging butterfly, and may not breed at all at Castle Hill. It is a pest of cultivated brassicas and, like the large white, is largely dependent on cultivated land for breeding. Clearly, in this case, the distribution of counts at a site is very much less informative than for the dingy skipper, and depends entirely on factors such as shelter from the wind and the presence of favoured flowers for adult feeding.

Unfortunately, the relationship between the distribution of adults, and of eggs and larvae has been studied for very few species. Even if there has been a study elsewhere, a relationship found at one site may not be fully applicable at another. For example, at one site the presence of flowers for adult feeding in areas suitable for oviposition may enable adults to remain in the breeding areas; at another site, flowers and larval food plants may be separate and the butterflies must then fly over a wider area. Such separation of breeding and feeding areas has been demonstrated for a population of the wood white in Sweden (Wiklund, 1977a), although in another study of the wood white, in England, there was no such separation (Warren, 1981).

Some butterflies show differences in mobility over their ranges. For example, Courtney (1980) suggests that the orange tip, which is wide ranging in the south of England, occurs in local sedentary populations in parts of the north. The holly blue may also be more sedentary in its univoltine populations in the north of England than in its bivoltine populations in the south; in the latter case the assumption of a difference in mobility is based on inference, with no direct evidence (Chapter 11).

5.3 EFFECTS OF SHADE

As butterflies fly by day and, on cool days, may require sunshine for flight, it is not surprising to find that shade, either by vegetation or by topographical features, is of particular importance to them. In woodlands, there is no doubt that the extent of shade has a major effect on butterfly distributions. Warren (1985) measured shade levels along butterfly transect routes in two woods, Yardley Chase in Northamptonshire and Monks Wood in Cambridgeshire. These shade measurements were made using vertical hemispherical photography (Figure 5.5), with a lens giving a 180 degree field of view. The relative abundance of butterflies was strongly related to shade, with the associations between individual species and particular shade levels usually repeated in both woods. Species found only in the most open

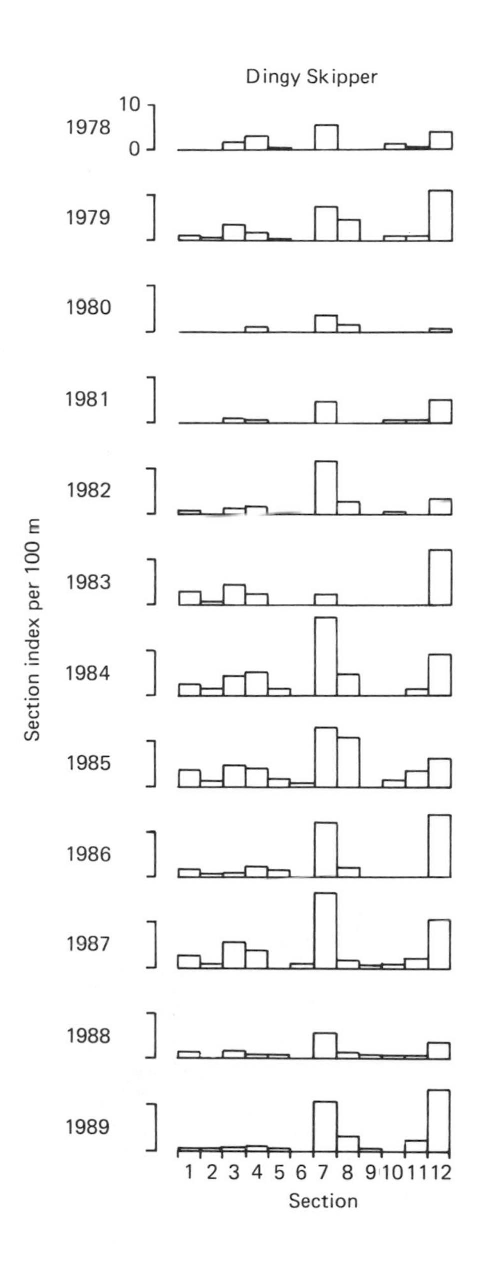

Figure 5.4 Distribution of section index values of the dingy skipper around the Castle Hill transect, 1978–89. Section index values corrected for section lengths. The distribution has remained similar over the recording period, with very few records in areas of improved grassland (sections 6 and 9).

situations were the grizzled skipper, dingy skipper, common blue and small heath, while the most shade-tolerant butterfly was the speckled wood, which was recorded at peak numbers at about 60% shade although it was tolerant of a wide range of conditions.

Greatorex-Davis *et al.* (1992) extended this work on shade in a study of transect counts in several conifer-plantation woodlands in the south of England. In 59 rides of these woods, with a very wide range of shade conditions, both the number of butterfly species recorded and the total number of butterflies of all species were strongly related to shade (Figure 5.6). Hall and Greatorex-Davis (unpublished report) examined the number of species of herbaceous plants in these same woodland rides and found a similar strong relationship with shade. It is likely that the association of a butterfly species with a particular shade level is not solely a direct response

Figure 5.5 Hemispherical photograph, using a lens with 180 degree field of view, taken vertically from the centre of a ride (track) in Monks Wood in Cambridgeshire. This type of photograph, with appropriate overlays, is used to assess the amount of potential sunlight which is interrupted by trees and other plants (direct shade) and also the extent to which all open sky is obscured (indirect shade).

to local microclimate, but is partly related to the presence of flowers for adult feeding and larval food plants for oviposition.

Elsewhere in the book, it has been emphasized that butterflies can fly in the absence of sunshine, provided that the shade temperature is sufficiently

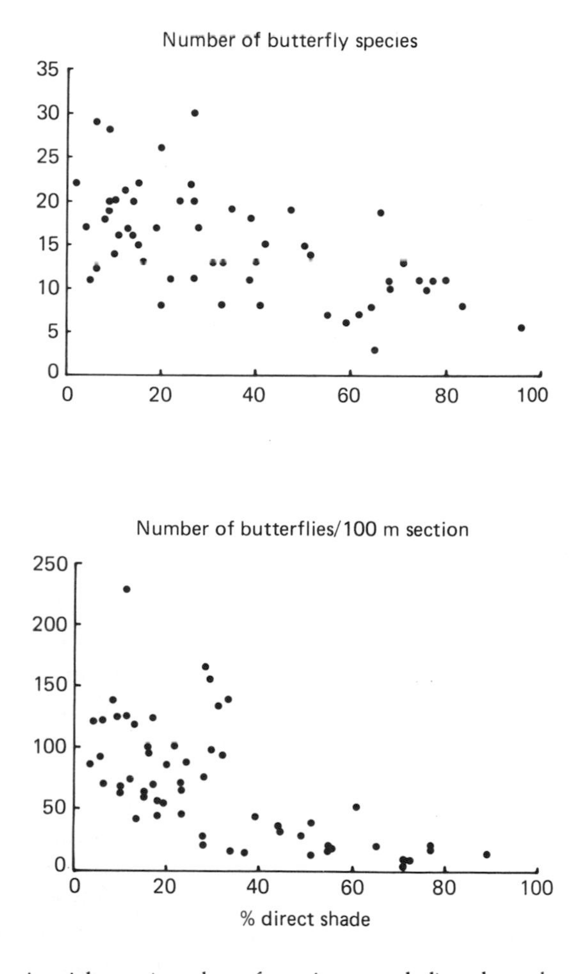

Figure 5.6 Species richness (number of species recorded) and number of individual butterflies recorded over a season, in relation to direct shade (percentage of potential sunlight intercepted by the tree canopy). Data for 59 sections of Butterfly Monitoring Scheme transects in seven plantation woods. The plantations are on ancient woodland sites in southern England. Many woodland butterflies are associated with open areas in woods, either recent clearings or, as in this case, open rides. Redrawn from Greatorex-Davis *et al.* (1992).

high. Nevertheless, sunlight is undoubtedly important in enabling flight when temperature is low and perhaps in speeding larval development. It is surprising therefore that Greatorex-Davis *et al.* (1992) found that the abundance of butterflies was more closely related to indirect shade (the proportion of the sky obscured by the tree canopy) than direct shade (potential sunlight intercepted by the tree canopy). The authors emphasize that their study did not relate to spring sunshine and suggest that it is in the cooler months that sunshine is likely to be of greatest importance. Thus the particular orientation of a ride may be important, as this has a major effect on the amount of sunshine reaching it.

Hall and Greatorex-Davis (unpublished report) found that the Heteroptera (true bugs) were also more abundant, in numbers of species and individuals, in open areas of woods. Undoubtedly, the general requirement of butterflies for open clearances within woodland is shared by many other insects, but there are also numerous exceptions, for example some insects are most abundant in densely-shaded woods (e.g. Sterling and Hambler, 1988), while others are entirely dependent on high forest (Fry and Lonsdale, 1991). Nevertheless, as a generalization, densely shaded woods are relatively numerous in Britain, while those managed by frequent, rotational, clearances are scarce and it is the butterflies, and probably other insects, that require such clearances that are most in decline.

5.4 SOME LOCAL DISTRIBUTION PATTERNS

In more open situations, where shade is of less overriding importance than in woodlands, local distributions of butterflies are likely to depend on factors such as shelter from the wind, due either to topography or low vegetation, and floristic differences between areas. Details of the structure and composition of vegetation are not generally available for sites in the monitoring scheme. Nevertheless, in many cases clear distribution patterns occur which can be related to environmental factors.

The mountain ringlet at Ben Lawers in the Eastern Highlands of Scotland is most abundant at between 700 and 1000 m altitude (Figure 5.7). Thomson (1980) writes that in Scotland this butterfly tends to be most common between 450 and 800 m, so the distribution at Ben Lawers is fairly typical. The food plant of the mountain ringlet is mat-grass, which is itself most abundant on mountains, although the extent to which the distribution of the butterfly is determined by the abundance of its food plant is not known. Mat-grass is favoured by sheep-grazing (e.g. Clapham *et al.*, 1962) and the likely reduction in numbers of sheep in marginal agricultural areas may have an impact on the mountain ringlet. However, insufficient is known of its requirements to predict whether or not the mountain ringlet will be adversely affected by such changes.

At Pewsey Downs in Wiltshire, the marsh fritillary had a very restricted distribution in the early 1980s. It was centred on an area from which grazing animals had been excluded for some 15 years. This lack of grazing appeared to have enabled the food plant, devils-bit scabious, to grow to the state favoured by the species. Porter (1981) has shown that large plants in exposed positions are preferred. From 1982 to 1984 the marsh fritillary was very abundant and spread to other parts of Pewsey Downs, although in later years it became scarce over the whole reserve. The marsh fritillary has long been known to undergo violent population fluctuations (Ford and Ford, 1930), perhaps associated with variable mortality caused by a parasitoid. In view of this, it would be unwise to look for the cause of the recent decline at Pewsey Downs solely in terms of vegetation changes.

Common farmland butterflies may sometimes show patterns of distribution just as striking as those of the rarer species. At Woodwalton Farm in Cambridgeshire, the transect route is largely along the edges of arable fields in which cereals are usually grown. These field edges generally have few butterflies (Figure 5.8), except in years when the small white is very abundant. Most species are concentrated along a green lane, which is bounded by hedges and scrub and has a central area which is usually (Chapter 4) rich in herbaceous plants. Part of the route is along the edge of a small wood, but the cereals abut almost on to the trees and the number of butterflies in this section is usually small.

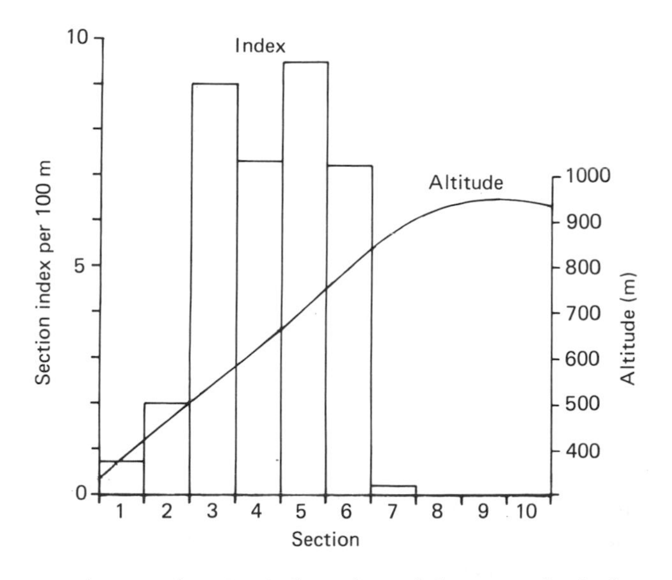

Figure 5.7 Distribution of section index values of the mountain ringlet at Ben Lawers in 1977. This is the only truly montane butterfly in Britain and is usually found at altitudes of 450–800 m. Its food plant is mat-grass, *Nardus stricta*.

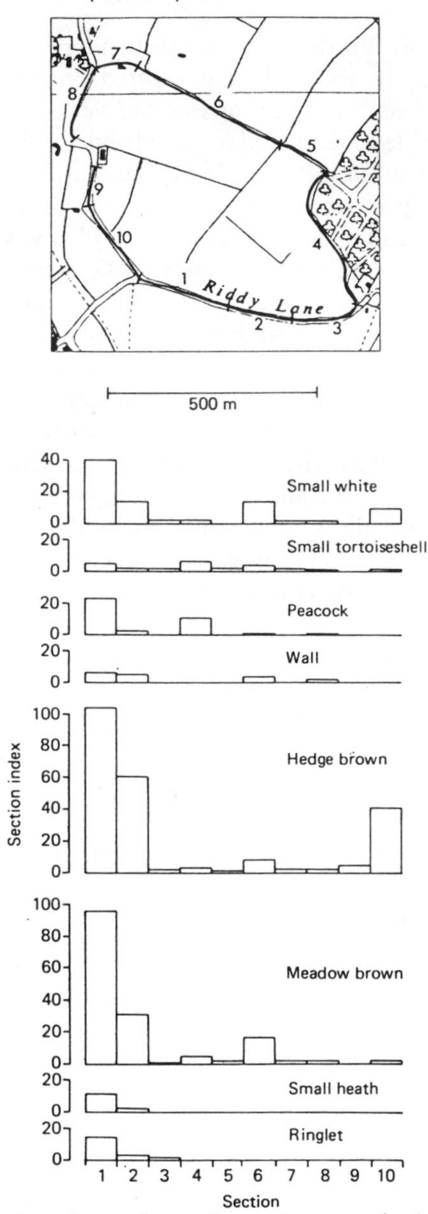

Figure 5.8 Distribution of some butterfly species around a farm transect in Cambridgeshire in 1980. Counts not corrected for section lengths. The importance to butterflies of the green lane is clear. Other butterflies recorded in that year were the small skipper, large skipper, brimstone, large white, green-veined white, orange tip, common blue, red admiral, painted lady and comma. Together with those figured, this is virtually the full complement of the widespread and common British butterflies.

5.5 STABILITY OF LOCAL DISTRIBUTIONS

A feature of the local distributions of butterflies at many sites is that they are often very stable from year to year. In some cases they have remained almost identical over more than a decade, as in the case of the dingy skipper at Castle Hill, discussed above (Figure 5.4). Further examples are given in the site studies (Chapter 13). Such stability is not surprising to the naturalist, who will know where to find, say, ringlets in a wood with which he or she is familiar and will be able to judge where they are likely to be found in unfamiliar woods in the same part of the country. The naturalist bases these decisions on experience, usually with little conscious assessment of the factors taken into consideration. However, the fact that such characteristic distributions occur and may be maintained for many years implies that there are specific conditions, related to topography, shade, shelter, floristic richness, larval food plants and perhaps other factors, which determine the patterns of distribution of butterflies.

Where there is considerable stability in the distribution of butterflies at a site, over a period of years, a corollary is that fluctuations in numbers in the different sections of the route tend to be synchronous over the period. Clearly, if numbers in different sections all fluctuated independently from generation to generation, any pattern in the distribution would soon be lost; in contrast, if there was perfect synchrony in generation to generation changes in numbers in different sections, the local distribution pattern would remain identical over the years. Between these extremes, which of course never occur in nature, there is likely to be enormous variation in the patterns of, and interdependence of, spatial and temporal distributions. This topic will be discussed in Chapter 14; for the present we emphasize the most general feature of the spatial distributions; that they tend to remain stable from generation to generation.

5.6 LOCAL DISTRIBUTIONS OF COMMON AND RARE SPECIES

It was suggested in Chapter 1 that the British butterflies divide sharply into the widespread species of the general countryside and the rare species which are, in the main, restricted to small, island biotopes within the general countryside. This restriction of rare species is largely because their special habitat requirements are met only in these 'islands', which are often nature reserves. This prompts the question, are nationally rare butterflies also more narrowly restricted on individual transect routes of the monitoring scheme than are the common species?

As yet, there has been no general study of monitoring scheme data to answer this question, but the Castle Hill results are used here for a first look at the topic. The dispersion of butterflies at Castle Hill, over a season (as in

Figure 5.2), has been assessed using an index of dispersion (Morisita, 1959). In this example, there is a clear relationship between national rarity and dispersion (Figure 5.9), with species that are rare nationally being also more restricted in their local distribution at Castle Hill than species that are widespread over the country.

A thorough study of this topic lies in the future, but it seems likely that the tendency for nationally rare species to have restricted local distributions will be found more generally. Two factors suggest that this may be so: the narrow habitat requirements of rare species and the fact that rare species are often very sedentary. In contrast, many of the common and widespread butterflies are very mobile and this mobility is likely to have contributed to more dispersed distributions.

5.7 DIFFERENCES IN ABUNDANCE FROM SITE TO SITE

The transect count method can be used for comparisons of the abundance of a species from site to site only with considerable care. One difficulty is that site differences may be partly confounded with recorder differences; another is that the transect routes may not adequately represent the sites through which they run. The following example will serve to illustrate other difficulties.

In 1989 the peacock had higher counts at sites in eastern England than in the south and southeast (Table 5.1), and this is also true of other recent years.

It is likely that the largely arable countryside of eastern England, with abundant nettles flourishing on nitrogen-rich soils, is particularly suitable for this nettle-feeding butterfly. However, a consideration of the biology of the peacock suggests that there are other possible explanations for the larger counts in eastern England.

The counts in the east were higher than in the south both in spring, after hibernation of the adult butterflies, and in summer, when the new generation emerges. In spring, males adopt vantage points from which to intercept females (Baker, 1972a). Woodlands are used particularly by the males in spring for their vantage points, although the females later move widely through the countryside for egg-laying. Woodlands occupy a much smaller proportion of the countryside in eastern England than in the south and the density of males in woods in eastern England may thus be greater, even if population levels in the surrounding countryside are no greater than in the south. In summer, peacock butterflies spend much of their time feeding at flowers before hibernation and, as in the spring, woodland sites are favoured. Suitable woods for flower-feeding are likely to be scarcer in eastern England and so these feeding concentrations of peacocks may be greater.

Table 5.1 Numbers of peacock butterflies recorded per site in each recording week in two regions of England in 1989. Counts were consistently higher in eastern England. The peacock probably breeds mainly in farmland where nettles are abundant, but uses semi-natural areas, especially woodland, for mating in the spring and feeding at flowers in the autumn. Semi-natural areas are scarcer in the east of England and so peacocks may concentrate their numbers in those areas that are available. Thus the larger counts in the monitored sites in the east may indicate greater density in the region as a whole but, alternatively, may reflect greater concentration in the smaller areas of woodland. (Note the earlier emergence of the autumn butterflies in the south)

Month	*Week*	*South and southeast England*	*East Midlands and East Anglia*	
April	1	1.4	5.8	
	2	1.1	2.9	Spring period of courtship, mating and egg-laying
	3	2.3	5.9	
	4	2.0	4.3	
May	5	2.3	3.7	
	6	1.3	2.8	
	7	0.7	1.9	
	8	0.5	1.0	
June	9	0.2	0.6	
	10	0	0.1	
	11	0.1	0.4	
	12	0.1	0.3	
	13	0	0	
July	14	0.1	0	
	15	1.6	0.3	
	16	5.5	8.6	Late summer period of feeding at flowers before hibernation
	17	9.5	26.7	
August	18	1.8	35.3	
	19	1.1	29.4	
	20	0.4	2.6	
	21	0.5	2.0	
	22	0.3	0.3	
September	23	0.3	0.7	
	24	0.3	0.2	
	25	0.6	0.2	
	26	0.4	0.6	

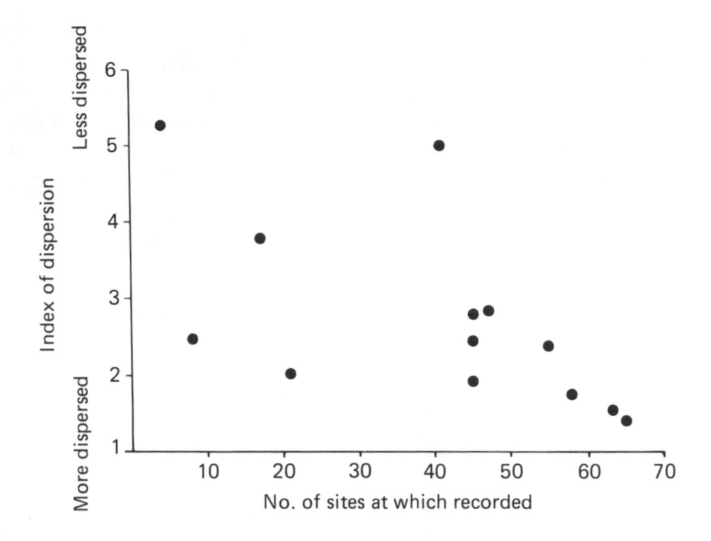

Figure 5.9 Relationship between dispersion of butterfly species at Castle Hill in 1981 and number of sites in the monitoring scheme at which the species were recorded in that year. (Spearman rank correlation $r = -0.72$, $P < 0.01$). Dispersion measured by Morisita's index of dispersion (Morisita, 1959). Eight species with very few records were omitted. There is a strong tendency for the rarer species to be less dispersed, i.e. concentrated into fewer sections; this may be because they have narrower habitat requirements than the common butterflies and also because they tend to be less mobile as adults.

It may be, and almost certainly is, the case that peacock butterflies are more abundant in eastern England than in the south, but this cannot be proved beyond doubt using the existing data from the Butterfly Monitoring Scheme alone. However, a fairly simple additional survey, conducted in one season using a random selection of sites as discussed in Chapter 3, could confirm whether there is a true difference in abundance in the two regions.

5.8 CONCLUSIONS

As far as we know, the Butterfly Monitoring Scheme has the only major body of data which records, in any detail, the local distribution of insects. In this chapter, examples have been given of some striking patterns of distribution and attention drawn, in particular, to the stability of many of these distributions from year to year. In later chapters, changes in local distributions are discussed in relation to changes in biotopes and more broadly in the context of the population ecology of butterflies.

The limitations of spatial data from transect counts, especially in comparing the abundance of butterflies from site to site, has been emphasized. In spite of these limitations, there is no doubt that these spatial data are of considerable interest, both in understanding of the population ecology of butterflies and for their use in detection of changes in their habitats.

– 6

Fluctuations in numbers

One of the most important generalisations that can be made about wild animal populations is that they fluctuate greatly in numbers.

Charles Elton, *The Ecology of Animals*, 1933.

6.1 INTRODUCTION

The main aim of monitoring is to record changes in abundance of butterflies, both at individual sites and more generally. Why is such information useful? What questions can it answer? Amongst the most obvious questions are:

1. Is there evidence of long-term trends in abundance, either increases or decreases, that may affect the overall status of butterfly species?
2. Are any rare butterflies of special conservation interest endangered by declines at individual sites?

Other, more general, questions include:

3. What is the typical character and magnitude of the population fluctuations of butterflies?
4. How do these fluctuations vary from species to species and over the country?

The first two questions have direct relevance to current decisions about the conservation of butterflies and about requirements for the more detailed research needed to understand the causes of changes. The second two questions may have little immediate relevance to the conservation of butterflies, but in attempting to answer them a start is made in the use of monitoring data to help to understand their population ecology. Monitoring cannot by itself supply such an understanding, but, together with more detailed studies, can make a major contribution. In the longer term, soundly based conservation policies depend on a broad knowledge of the population ecology of butterflies. At the end of this chapter, we return to these questions and consider the extent to which they have been answered.

6.2 SHORT-TERM FLUCTUATIONS AND LONG-TERM TRENDS IN NUMBERS OF THE COMMONER BUTTERFLIES

This section is based on 29 of the commoner butterfly species, listed in Table 6.1, for which there are sufficient data for the calculation of collated index values based on all sites. Most of these species are widespread and are generally common in the countryside, especially in southern Britain (Chapter 1). However, a few, such as the dingy skipper and white admiral, are amongst those which are restricted to 'island' biotopes within the agricultural countryside.

6.2.1 Calculation of collated index values from individual site data

Regional and all-sites changes in abundance have been calculated using the method of ratio-estimates (Cochran, 1963). The ratio-estimate is based on index values at all sites which have been recorded in successive pairs of years, and is simply the sum of the index values in year 2 divided by the sum of index values in year 1. The collated index values are compiled from an arbitrary 1976 index (a value of 100 was chosen), and multiplying this first value by successive ratio-estimates. Thus, the collated index values show only changes relative to the initial 1976 value.

As with the site index values, there is one collated index each year for univoltine butterflies (one generation each year) and also for those with overlapping flight periods or otherwise complex voltinism (Chapter 2). For species which are usually bivoltine there are two collated index values each year. In general, collated indexes have been calculated only if data from seven or more sites are available in every year. Usually, the number of sites is much larger than this.

The use of ratio-estimates is not the only possible method of calculating the collated index values. However, it has the advantage of simplicity; thus, on the basis of one count in each recording week at each site, the ratio-estimate is derived directly from the total number of butterflies recorded at all sites in successive generations. However, an assumption, inherent in the use of ratio-estimates, is that each site represents a sample drawn from a single nationwide population of butterflies. This may be more or less true for a few very wide-ranging species, but in many cases, the counts at individual sites represent discrete populations. In such cases, the use of ratio-estimates is convenient, but not strictly appropriate.

The method of calculating these, and most other, collated index values has the property that any errors or aberrations in the data for a site in one year continue to affect the index in all subsequent years (Moss, 1985;

Greenwood, 1989). If the site index values introducing such an aberration are large, this is potentially a serious problem, although we believe that the butterfly data are not badly affected in this way.

Alternative methods of collating the index values, which reduce this potential problem, are being tested and generally show only minor differences from the values obtained using ratio-estimates (Moss and Pollard, 1993). However, it is possible that one of these methods may eventually be preferred.

6.2.2 General characteristics of fluctuations in index values

Collated index values have been calculated for the 29 commoner butterflies, using data from all of the sites where they have been recorded. One of these 'species' is in fact a combination of two species, as the data for the small and Essex skipper are combined.

Fluctuations in collated index values (Figure 6.1) showed a wide variety of patterns. These ranged from the violent fluctuations in the index of the painted lady (overall range of index values × 66) to the relative stability of the small heath (range × 2). The data for the painted lady resemble random numbers, in that there was no significant relationship between index values in successive years.* The painted lady is migratory, on a trans-continental scale (Chapter 9), and the British contribution to overall numbers of this species is likely to be negligible. In addition, there seems to be a large element of chance in the number of individuals that reach Britain each year. Thus the numbers that arrive in a given year are likely to be, at most, weakly related to numbers in Britain in the previous year. In other cases, such as the marbled white,† there was a clear relationship between the index for one year and that of the previous year; such a relationship suggests that the species takes more than one year to recover from low population levels or to decline from high ones.

For bivoltine species, there was a within-season pattern in addition to the year-to-year fluctuations. In these species, the first generation index was usually smaller than the second generation index of the same year. This is particularly striking in the case of the wall, giving the graph a 'saw-tooth' appearance (Figure 6.1). The inference from this pattern is that the breeding success of the spring adults is consistently greater than that of the second generation, the larvae of which overwinter. A small third generation may occur in some species in some years; individuals from this third generation are included with the second generation index and strengthen this saw-tooth pattern a little. In the case of the wall, however, the third generation usually

* In fact, the correlation was negative, $r = -0.32$.
† $r = 0.66$, $P < 0.05$.

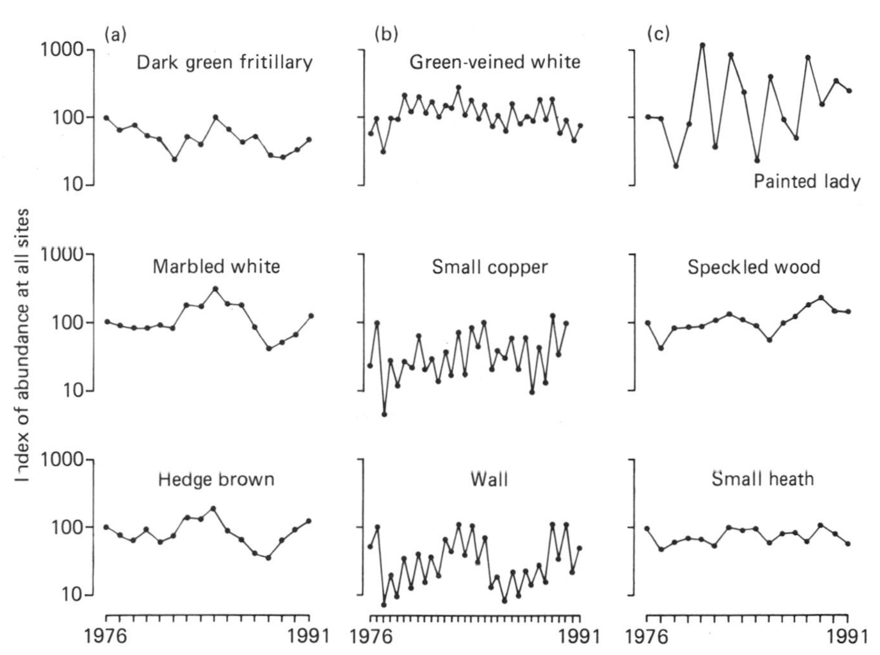

Figure 6.1 Fluctuations in abundance of nine butterfly species, 1976–91, based on collated index values from all sites in the monitoring scheme. (a) Species with one generation a year. (b) Species with two distinct generations. (c) Species with overlapping generations, given a single annual index. Note the logarithmic scale in this and subsequent figures of population fluctuations.

flies in October, after the monitoring period has ended.

There is a tendency for the fluctuations of different species to be correlated (using the second generation of bivoltine species). In the matrix of 36 correlations between species in the examples illustrated (Figure 6.1), there were nine (25%) significant positive correlations and one (3%) significant negative correlation (that between the dark green fritillary and speckled wood).* For the complete matrix of 29 species the percentages were similar, with 21% significant positive correlations and 2% negative. The fluctuations of the hedge brown (Figure 6.1) were significantly correlated with those of eight other species, and can be considered representative of one common pattern of fluctuations.

6.2.3 Variability of fluctuations

The generalization made by Elton (1933), quoted at the start of this chapter, that variability of animal populations is typically very large, was part of a general discussion of population fluctuations. Elton was countering the

* $P < 0.05$.

nineteenth century concept of a 'balance of nature', in which populations were assumed to be more or less constant, apart from disturbances due to interference by humans. Since Elton wrote this in 1933, population variability has continued to interest ecologists and provide material for discussion and argument right up to the present day. Do fluctuations in butterfly numbers support Elton's generalization? Perhaps the most marked feature of the fluctuations of these 29 butterfly species is the contrast between species; some fluctuate wildly, some seem to be remarkably stable.

A measure of variability is the geometric mean change (increase or decrease) in population size from generation to generation (as used by Pollard, 1984); this was used here although different measures have been adopted in other studies (review by McArdle *et al.*, 1990). For the most variable species, the painted lady, the 1976–91 geometric mean change from one year to the next was × 6.0; the least variable was the meadow brown (× 1.2).

The 16 univoltine species all had remarkably similar, and relatively small, variability, lying between × 1.2 for the meadow brown and × 1.6 for the white admiral. As a group they were more stable than the bivoltine species, with little overlap. Two of the 16 species, the peacock and brimstone, have one generation but two flight-periods; the adults overwinter and the same individuals fly in the autumn and spring. The variability of these species, calculated using only the flight-period before hibernation, was similar to that of other univoltine species and the brimstone was amongst the most stable (× 1.3). Average changes of this order seem to be very small when the potential for change in butterfly populations is considered. Butterflies lay upwards of 200 eggs and so have enormous potential for increase, while catastrophic mortalities, which would seem to be an ever-present threat, also seem to be rarely realized. However, changes of the order of × 1.5, if continued for several years in the same direction, soon result in very large changes in abundance.

The variability of the bivoltine butterflies was measured using the second index in each year, as with the univoltine peacock and brimstone. As a group the species with two or more generations a year were more variable than the univoltine species, with little overlap. The range was from × 1.5 for the green-veined white to × 3.6 for the holly blue. This contrast in variability between univoltine and bivoltine species cuts across taxonomic groupings. The brimstone and orange tip, the univoltine periods, were the least variable of this family and the univoltine green hairstreak was the least variable of the lycaenids; the wall, the only bivoltine satyrid, was the most variable of this family. Thus, the greater variability seems likely to be directly related to voltinism. Additional generations obviously add to the potential for rapid increase in population size and a corollary is the additional risk of large mortality in vulnerable stages in the life cycle.

Within the bivoltine butterflies, the holly blue was much the most variable; such large variability is likely to be associated with frequent extinction and re-establishment of local populations, as appears to be the case with the holly blue (Chapter 7).

A third group of butterflies (Table 6.1) have rather complex life cycles and we use a single annual index value, rather than attempt to separate those for the different generations. It is difficult, therefore, to relate their variability to that of the butterflies which are given an index for each generation. This third group includes the most variable species, the painted lady, the other common migrant, the red admiral (which is much less variable than the painted lady) and also one of the most stable species, the small heath (× 1.3).

Simple inspection of graphs of population fluctuations of the meadow brown in different parts of Britain (Pollard *et al.*, 1986) suggested that northern populations were much more variable than those in the south. A general analysis of geographical trends in variability is in progress (Thomas, Moss and Pollard, unpublished) and early results support this impression. This analysis suggests that populations of several species tend to be more variable either towards the north or towards the east of the country. Interpretation of these trends in variability is not straightforward, particularly as variability is related to abundance, which may also vary in different parts of the country. If the geographical trends in variability are confirmed, they have clear implications for the likely frequency of extinction of local butterfly populations in different parts of the country. As a generalization, highly variable populations are more likely to be subject to local extinctions although other factors may also be relevant (Chapter 7).

6.2.4 Longer-term trends

The graphs in Figure 6.1 suggest that some species have changed in general levels of abundance in the recording period; for example, the index values for the dark green fritillary indicate that a decline may have occurred and those of the speckled wood suggest an increase. Changes in abundance may be gradual or sudden, for example through a sharp increase or decline in one year which is maintained in subsequent years, or a more complex mixture of effects. Because population data are time-series, in which the population in one year is expected to be related to that in the previous year, simple trends with a scatter of points about a trend line are not likely to occur. Inspection of the data (Figure 6.1) does indeed show a variety of complex patterns. In addition, if the method of calculating collated index values does introduce biases, a possibility raised earlier in this chapter, it is the long-term trends which are most likely to be affected. In spite of these potential difficulties, examination of trends in collated index values* over the recording period has been used here as a first step in identifying changes in abundance (Table 6.1). As a check on the results from all-sites indexes, significant trends at individual sites have also been calculated (Table 6.1). Zero values were excluded from these calculations, so that these trends at individual sites are distinct from the population extinctions and colonizations considered later (Chapter 7; Table 7.1). If both the all-sites indexes and individual site data indicate significant trends in abundance, this is good evidence that the trends are real.

The species with all-sites index values which have increased significantly are the ringlet, peacock, small copper, comma and speckled wood (Table 6.1). All of these species except the small copper have expanded their ranges during the twentieth century and, in most cases, during the last 20 years (Chapter 1). The comma and speckled wood have colonized new sites in the monitoring scheme (Chapter 7).

The butterflies for which all-sites collated index values are calculated and which have declined significantly in numbers, are the dingy skipper, grizzled skipper, dark green fritillary and grayling; because of the relative rarity of all of these species, the evidence for decreased abundance is based on few sites and is therefore less reliable than for the increasing species. In particular, the data for the grayling have been strongly influenced by the very large population at Tentsmuir Point.

The individual site data (Table 6.1) confirm the increase in abundance of the ringlet, small copper, comma and speckled wood, and provide some support for increase in abundance of the peacock. In addition the meadow brown has increased in numbers at many sites in eastern England, but this is not reflected in the all-sites index because of declines at sites elsewhere in the country. The decrease of the dark green fritillary, shown by the all-sites indexes, is given some additional support by the individual site results, but the data are few and the trend remains in doubt.

The general trends in abundance identified are consistent with the changes in status outlined in Chapter 1, with some common species increasing in abundance and expanding their ranges and relatively rare species tending to become rarer. If the species are divided according to status, as was done in Chapter 1, the rare species show a significantly higher proportion of declines than do the common species (Table 6.2).

The speckled wood has colonized new sites in the monitoring scheme in eastern England (Chapter 7). At individual sites, where it has long-standing resident populations in the east, there have been very large increases in numbers. In contrast, the hedge brown, which has spread north across the whole of its range, shows no relative increase in any region. Thus there is no

* Log_{10} transformation.

Table 6.1 Trends in the abundance† of 29 butterfly species from 1976 to 1991. The results are presented in two ways. (1) Trends in collated index values, based on all of the sites in the scheme: significant long-term *increases and †decreases. (2) The number of significant ($P < 0.05$) trends at individual sites with data for at least 8 years. Zero index values were not included. The species which have increased tend to be more widespread (present at more sites) and the data therefore more conclusive than for those which have decreased

	Trend in all-sites index	*Significant trends at individual sites*		*No. of individual sites tested*
		+	–	
Univoltine species				
Small skipper	0.014	6	6	(54)
Large skipper	0.017	8	4	(53)
Dingy skipper	–0.019†	0	2	(22)
Grizzled skipper	–0.032†	0	0	(10)
Brimstone	0.002	4	2	(39)
Orange tip	0.008	9	3	(55)
Green hairstreak	–0.003	1	4	(21)
White admiral	0.001	0	1	(10)
Peacock	0.021*	9	6	(64)
Dark green fritillary	–0.021†	1	5	(23)
Silver-washed fritillary	0.004	0	1	(8)
Marbled white	–0.006	5	2	(21)
Grayling	–0.023†	0	2	(18)
Hedge brown	–0.005	5	9	(58)
Meadow brown	0.009	17	7	(80)
Ringlet	0.052*	25	0	(47)
Bivoltine species				
Large white	–0.010	6	5	(67)
Small white	–0.014	5	5	(68)
Green-veined white	–0.008	6	7	(65)
Small copper	0.025*	7	0	(54)
Common blue	0.021	3	1	(45)
Holly blue	0.007	2	0	(10)
Wall	0.004	1	3	(42)
Ill-defined flight-periods (annual changes)				
Red admiral	0.034	7	2	(59)
Painted lady	0.032	1	0	(29)
Small tortoiseshell	0.024	5	3	(74)
Comma	0.048*	10	1	(43)
Speckled wood	0.024*	10	0	(42)
Small heath	0.005	7	10	(62)

† Regression of $\log_{10}$ index on years

Table 6.2 Significant trends in index values of butterflies at individual sites, 1976–91. Species divided into widespread, common species and localized, rare species (see Chapter 1 for a list of species in each category). The common species, as a group, have been more successful than the rare species (chi-squared = 10.1, $P < 0.01$)

	Positive trends	*Negative trends*	*No. of populations tested*
Widespread, common species	155	74	1121
Localized, rare species	16	24	210

consistent association between increase in abundance over the recording period and expansion of range.

6.2.5 Synchrony of changes

One striking feature of the monitoring data has been the synchrony of population changes from site to site, over regions of Britain, and sometimes throughout the range of a species. Such synchrony is true not only of wide-ranging butterflies, such as the peacock and small tortoiseshell (Pollard *et al.*, 1986), but also of more sedentary butterflies that are considered to form discrete local populations. The synchrony has been demonstrated (Pollard, 1991a) by using data for six species, that are generally considered to be sedentary, from five sites. The species were the large skipper, small copper, common blue, hedge brown, meadow brown and ringlet. The sites were those which have been recorded by the same person throughout (eliminating one source of potential variability), with the addition of Monks Wood. The sites include woodlands, heaths and downs on soils ranging from heavy clays, to acid sands and chalk.

The extent of the synchrony is illustrated here in two ways. First, trends for four of the species considered by Pollard (1991a) which were present in adequate numbers at Monks Wood were compared with collated trends at all sites in the scheme (Figure 6.2). In each case, it is clear that the fluctuations in index values at Monks Wood are dominated by factors which operate widely over Britain. In the cases of the large skipper and ringlet there are indications that these species may have been faring better at Monks Wood than at most sites, but even in these cases the site data are closely correlated with the collated index values.

Monks Wood is located near the 'centre' of the monitoring scheme sites (given that there is a disproportionately large number of sites in the south of

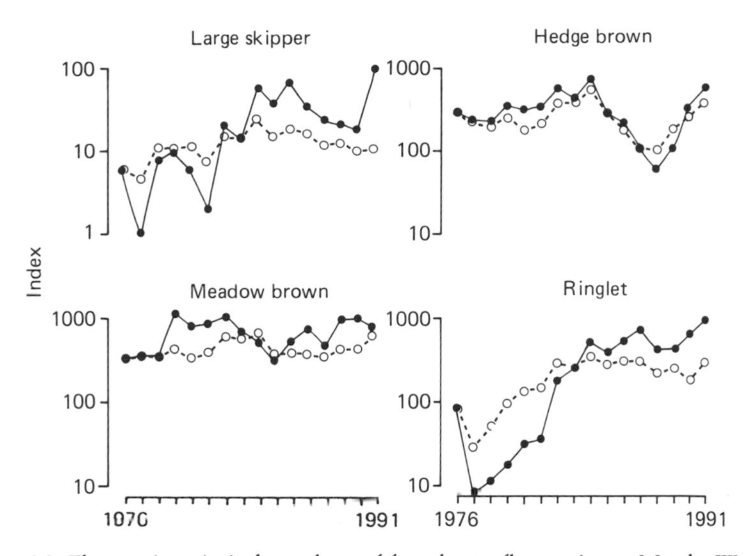

Figure 6.2 Fluctuations in index values of four butterfly species at Monks Wood (solid line), compared with fluctuations at all sites (excluding Monks Wood) (dotted line), 1976–91. The 1976 all-sites index is given the same value as that at Monks Wood. The general synchrony of fluctuations suggests that a widespread factor, presumably weather, has a major influence on the local populations. There is an indication that both the large skipper and ringlet are increasing relative to the all-sites indexes.

England). Thus, on the assumption that weather has an important influence on fluctuations, those at Monks Wood may perhaps be expected to approximate to some sort of average of sites in the scheme. However, the synchrony is found at widely scattered sites, as illustrated by data for one species, the hedge brown (Figure 6.3). Considering the geographic range of the sites, from Studland Heath in Dorset, on the south coast, to Walberswick in Suffolk, near to the coast of East Anglia, such close synchrony of populations over a long period is remarkable. While the similarity in fluctuations of hedge brown over the country is particularly striking, the same tendency is evident in most species; further examples are given in the site studies (Chapter 13).

There are, however, a few exceptions to this general synchrony of population fluctuations and it tends to break down at sites farther north and west. For example, the peacock, brimstone and small tortoiseshell became relatively scarce towards the north of their ranges in the mid-1980s (e.g. Wilson, 1991), but there was no equivalent decline in the south. Similarly, the recent sharp increase in numbers of the holly blue very clearly began in the extreme south, and only extended to more northern sites a year or two

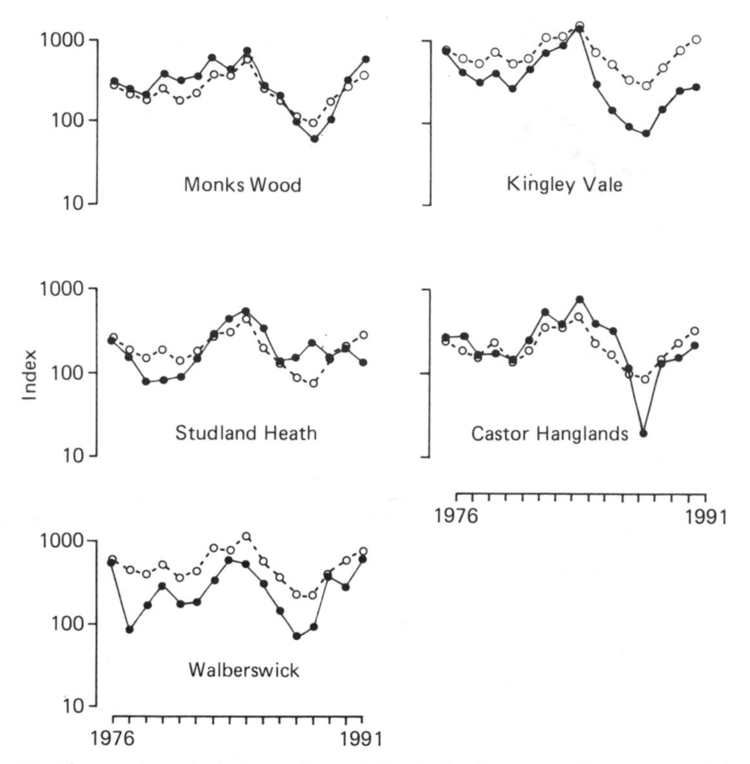

Figure 6.3 Fluctuations in index values of the hedge brown at five sites (solid lines), compared with fluctuations at all sites (excluding the particular site in each graph) (dotted line). As in Figure 6.2, the general synchrony, at widely separated sites, suggests a strong effect of weather.

later (Chapter 11). The wall has also undergone locally asynchronous population fluctuations.

Some synchrony of butterfly populations is to be expected, and it would be surprising if it did not occur. If, for example, the speckled wood is very abundant in a particular year in Oxfordshire, we would also expect it to be abundant in the neighbouring counties of Berkshire and Buckinghamshire. Neither is it surprising that wide-ranging butterflies, such as the large white, show synchrony of changes, as the population units are very large. However, close synchrony, maintained over many generations, of widely separated populations of species that are generally considered to be sedentary, such as the hedge brown and ringlet (e.g. Thomas and Lewington, 1991), is one of the most unexpected features of the monitoring data. Presumably, widely separated and discrete populations are exposed to a considerable variety of different predators and competing herbivores, but such factors seem to have little effect on population fluctuations; certainly they do not determine their general character.

The obvious conclusion to be drawn from the general synchrony of population fluctuations of individual butterfly species is that the fluctuations are driven by a factor which operates over much or all of Britain. One such widespread factor is weather, and the possible effects of variability in weather on butterfly populations is discussed in some detail in Chapter 8.

6.3 POPULATION FLUCTUATIONS OF THE RARER SPECIES

In addition to the 29 species for which collated index values are produced, there are a number of more locally distributed or rare butterflies for which there are long datasets at individual sites. A few examples (Figure 6.4) show a similar range in patterns of fluctuations to those of the commoner species. The brown argus at Swanage has shown the large fluctuations characteristic of bivoltine butterflies, although in this case with no clear disparity in size between first and second generation indexes; the high brown fritillary and Scotch argus are, in contrast and like other univoltine species, much less variable.

Because they are present at few sites, we can make little use of the synoptic nature of the monitoring scheme for these rarer species. As more data are collected, it will be possible to be more ambitious; for example, to compare at the same sites fluctuations of a northern butterfly such as the Scotch argus with those of a similar, but ubiquitous, butterfly such as the meadow brown. In this way we hope to learn something of the population ecology of a northern butterfly and perhaps of adaptation to a harsh climate.

For the present, the main interest in the data on changes in numbers of the rarer species lies in the detection of trends in numbers at individual sites which may indicate that conditions are deteriorating. There have, unfortunately, been a few examples in the monitoring scheme, in which rare butterflies have declined in numbers and are now, apparently, extinct. Equally, however, there have been examples of management for rare butterflies that have been followed by sharp increases in index values. Instances of such 'failures' and 'successes' feature in the accounts of rare butterflies and the site studies in Chapters 12 and 13, respectively.

6.4 CONCLUSIONS

As might be expected, the four questions posed at the beginning of this chapter have been answered only incompletely. Evidence of long-term trends – mostly increases in abundance accompanied by expansions of range – has emerged. In particular, the common butterflies have fared relatively well at sites in the scheme since 1976, while in contrast the rare

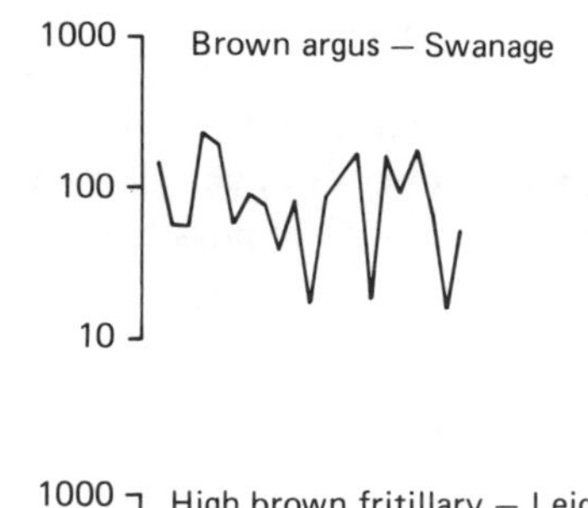

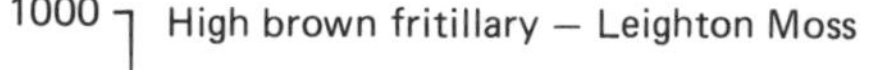

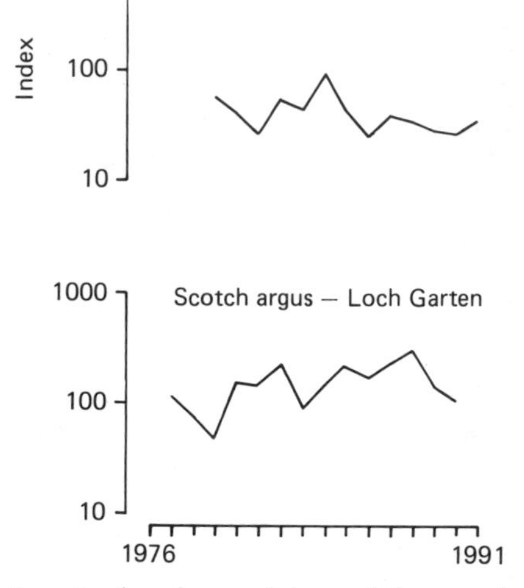

Figure 6.4 Fluctuations in abundance of three of the rarer butterflies at individual sites in the monitoring scheme. The brown argus shows the wide fluctuations typical of butterflies with two generations each year, while the univoltine species are, characteristically, more stable.

species show a predominance of declines. As has been discussed earlier, it is possible that the sites in the scheme present a more favourable picture than would more representative sites.

Progress has been greatest on the more general aspects of the population ecology of butterflies. The monitoring scheme has provided results which add a new dimension to knowledge of the population ecology of butterflies. A striking finding has been the pronounced synchrony of butterfly populations over large areas of the country. The implications of this synchrony are discussed further in Chapter 8 (effects of weather) and in Chapter 14 (population ecology).

7

Colonization and extinction

... my friend, Mr. R.F. Bretherton published in 1939 an exceptionally thorough survey of the butterflies of the Oxford district. At that time neither he nor I had seen the Speckled Wood in Bagley during 19 years collecting in that forest and Mr. Bretherton was able to obtain only one record of its occurrence (in 1922). In the year 1943 I saw six specimens there in a single afternoon and it is now appearing even in the city of Oxford....

E.B. Ford, *Butterflies*, 1945.

7.1 INTRODUCTION

It is an exciting moment when a naturalist records a new species in an area he or she knows well and equally depressing to become aware that a familiar butterfly has disappeared. For the population ecologist the study of colonization and local extinction is fascinating, but very often frustrating. A knowledge of the frequency of establishment and loss of populations is one of the keys to an understanding of population processes; do populations persist for tens or hundreds of generations, with colonization and extinction only rare events or, alternatively, are there frequent extinctions and foundations of local populations even in the heart of a species' range? If the answers to such questions are known, conservation measures can be planned more effectively. If, for example, a butterfly species is known to undergo frequent local extinctions, apparently unrelated to changes in its habitat, there is little point in establishing a single small nature reserve for its conservation.

Strictly, recorded extinction (or colonization) in the monitoring scheme applies to the transect route rather than to the whole site, although in many cases the two can be equated. The records of the white-letter hairstreak in Monks Wood provide one example in which the transect route cannot be equated with the site as a whole. In Monks Wood, the white-letter hairstreak was recorded in each of the 7 years from 1973 to 1979. Its food plants, elms, were dying of Dutch-elm disease during this period and few, if any, mature trees survived along the transect route. In 1979, the disease was particularly severe and the last trees were dying during the flight-period of the butterflies. In spite of this, many more white-letter hairstreaks were seen than in any of the previous years. This butterfly usually flies high in the canopy of trees (not always elms) and is seen only occasionally at ground level; it seemed likely that in 1979 the butterflies were dispersing widely in a vain search for elm trees. From 1980 to 1990, no white-letter hairstreaks were seen during counts and, even though it is an inconspicuous butterfly, this seemed to be reliable evidence of extinction. However, it had survived, together with a few elms, in another part of the wood (Welch, 1989) and a single individual was seen on the transect route in 1991.

The frustration in studying colonization and extinction lies in this difficulty in proving that a butterfly is absent from an area. If a species is not recorded for a period of several years, and is then seen again, as in the case of the speckled wood quoted at the beginning of the chapter, what can be concluded? Was it present in Bagley Wood throughout the 1920s and 1930s, but in such low numbers that the chance of a sighting was remote, or was it absent for much of the period with recolonization only in the early 1940s? It seems reasonably certain that the speckled wood recolonized Bagley Wood after earlier extinction, but a small element of doubt always remains in such circumstances.

To try to cope with these difficulties in the study of the extinction and foundation of local populations in the Butterfly Monitoring Scheme, arbitrary definitions have been used to decide whether to assume that a butterfly species was present at or absent from a site. In this chapter, some results from an analysis based on such assumptions are described (Pollard and Yates, 1992) and additional types of evidence, present in the monitoring data, for extinction and colonization are considered. To avoid tedious repetition, identified extinctions and colonizations are referred to without the qualifications 'identified' or 'probable', but these qualifications are always implied.

7.2 IDENTIFICATION OF EXTINCTIONS AND COLONIZATIONS IN MONITORING DATA

The preliminary study of extinctions and colonizations in the monitoring data (Pollard and Yates, 1992) was restricted to the commoner butterflies for which collated index values have been calculated, but here results for the rare butterflies are also given. Two common migratory butterflies, the red admiral and painted lady, were excluded because, in effect, they became extinct in every year and recolonize the country from abroad (although a small number of red admirals overwinter in Britain; see Chapter 9).

To identify extinctions and colonizations we first needed to identify assumed 'presence' or 'absence'.

1. It was assumed that a breeding population was present if a butterfly species was recorded in four successive flight-periods and absent if there were no records for four successive flight-periods. Other combinations of records were considered inconclusive.
2. Extinction at a site was assumed if 'presence' was followed by 'absence' at some later time.
3. Colonization was assumed if 'absence' was followed by 'presence'.

The above definitions are by no means the only possibilities and were adopted because they seemed reasonable and gave us a lead to an initial study. Using these arbitrary definitions, 123 colonizations and 79 extinctions were identified in the period 1976–91 (Table 7.1). It is likely that many of these identifications were of true colonizations and extinctions, but some were undoubtedly misidentifications, for example because a species was present and breeding at a site in small numbers, but was not seen on transect counts and was classified as 'extinct'.

Extinctions and colonizations may occur within the range of a species or may be a consequence of changes of range. In the latter case, a predominance of either extinctions or colonizations would be expected. In the monitoring data (Table 7.1) there was a general correlation between the numbers of extinctions and colonizations, suggesting that in the main they relate to losses and gains of populations within the ranges of species.

7.3 EXTINCTION AND COLONIZATION WITHIN THE RANGE

For species with relatively large numbers of both extinctions and colonizations, such as the holly blue, wall, small copper and common blue, it is likely that these reflect a genuine high turnover of local populations within the range of the species. On the other hand, the absence of extinctions and colonizations is less conclusive. Absence of identifications may indeed reflect a small turnover of local populations, but it is also likely that the mobility of some butterflies is such that individuals, from elsewhere in the general locality, fly into a study site before extinctions can be identified. The available information on the mobility of butterflies (Appendix A) suggests that the grayling, for example, probably falls into the first category, with correct identification of little change in the number of populations, and the small tortoiseshell into the second. The small tortoiseshell could fail to breed at an individual site for many generations, but the mobility of this species is such that local extinction could easily remain unidentified; it is a wide-ranging butterfly and adults from breeding populations elsewhere are

Table 7.1 Identified extinctions and colonizations of butterfly populations at sites in the Butterfly Monitoring Scheme, 1976–91. The number of sites are those at which the species has been recorded and which have sufficient data for tests. The tests for extinction and colonization are based on presence and absence for a specified number of flight-periods (see text) and for this reason there are likely to be more identifications for species with two flight-periods each year. As the identifications depend upon strict 'rules', some assumed extinctions and colonizations mentioned in the text are not included in this table. Data in part from Pollard and Yates (1992), but the definition of sites for inclusion differs slightly

	Sites	*Extinctions*	*Colonizations*
Univoltine species			
Small skipper	57	0	3
Lulworth skipper	1	0	0
Silver-spotted skipper	4	0	2
Large skipper	54	1	5
Dingy skipper	22	0	2
Grizzled skipper	10	1	0
Swallowtail	1	0	0
Wood white*	4	1	1
Brimstone	53	3	2
Orange tip	56	1	1
Green hairstreak	22	1	1
Small blue	5	0	0
Northern brown argus	3	0	0
Silver-studded blue	1	0	0
Common blue (northern)	11	0	0
Chalkhill blue	9	0	1
Duke of Burgundy	2	0	0
White admiral	14	2	5
Peacock	76	6	3
Small pearl-bordered fritillary	13	2	2
Pearl-bordered fritillary	10	0	2
High brown fritillary	4	1	0
Dark green fritillary	27	2	1
Silver-washed fritillary	11	0	0
Marsh fritillary	4	1	0
Heath fritillary	2	0	0
Scotch argus	2	0	0
Marbled white	22	0	1
Grayling	19	0	1
Hedge brown	60	0	0
Meadow brown	80	0	0
Large heath	2	0	0
Ringlet	47	0	0

Bivoltine species			
Large white	84	4	0
Small white	82	2	1
Green-veined white	88	1	4
Small copper	56	8	9
Common blue	61	7	11
Adonis blue*	4	1	2
Holly blue	25	17	39
Brown argus	15	2	1
Wall	57	15	14
Indistinct voltinism			
Small tortoiseshell	81	0	0
Comma	43	0	7
Speckled wood	41	0	3
Small heath	65	0	1

*The Adonis blue and wood white include introduced populations.

likely to be seen almost anywhere in most years, and will breed immediately suitable conditions recur.

The holly blue, and to a lesser extent the wall, account for a large proportion of extinctions and colonizations. The frequency of these events, per site, was very much higher for the holly blue than for any other butterfly; for example, at 15 individual sites, both extinction and colonization were recorded. It seems that the population dynamics of this species are atypical of British butterflies in general, as is discussed in more detail in Chapter 11.

There have been several studies of relationships between the frequency of extinction and colonization and other characteristics of life cycles and fluctuations in abundance (e.g. Hanski, 1982; Pimm *et al.*, 1988). These topics have been touched on by Pollard and Yates (1992), although the data from the scheme are only just becoming sufficient for such a study. A tentative conclusion is that frequent local extinction is most likely in species which undergo periods of general national scarcity lasting for several generations. As is described in Chapter 8, these periods of scarcity are, in general, related to unfavourable weather. In the 1976–91 period, the species with highest turnover of populations were all bivoltine (even taking into account the longer datasets); whether this is a matter of chance or is a characteristic feature of bivoltine species is not yet clear.

7.4 EXPANSIONS AND CONTRACTIONS OF RANGE

Changes in range of butterflies obviously involve extinctions or colonizations at the edge of the range. In recent decades, many of our commoner butterflies have extended their ranges and some rare species have become even more restricted. Both of these trends have been reflected in gains and losses of populations at monitoring scheme sites.

Of the butterflies which have extended their ranges in recent years (Chapter 1), the small skipper, comma and speckled wood have colonized monitored sites at the edges of their ranges. In other cases, such as the hedge brown, ringlet and orange tip, there are few monitored sites in the areas of expansion. For the comma there were no extinctions, but there were seven colonizations (Table 7.1), six of which were close to the northern edge of the present range (Figure 7.1). The spread of the comma has continued, albeit with temporary reversals, over many decades (Pratt, 1986–87) and the data from the monitoring scheme suggest that it is still in progress.

As the quotation at the start of the chapter indicates, the spread of the speckled wood, like that of the comma, began long ago (e.g. Downes, 1948; Chalmers-Hunt and Owen, 1952). Also like the comma, the range of the speckled wood contracted in the last century and the recovery is of some, but not yet all, of this lost ground. At the time of publication of Heath *et al.*'s (1984) distribution maps, it was still generally scarce or absent in large parts of eastern England, apart from the area of Thetford Forest; subsequently, the edge of the main range has pushed eastward and populations have become established at the monitored sites of Potton Wood, Wicken Fen, Woodwalton Fen and Chippenham Fen (Figure 7.2). Heath *et al.*'s (1984) map shows a gap in the distribution of the speckled wood in the London area, but Plant (1987) comments that it is 'encroaching upon the capital principally from the Surrey and Kent side'. At the monitored site of Hampstead Heath, there were intermittent records in the early 1980s, but it has apparently become firmly established and increased in abundance only in the last 3 years.

The search of the monitoring scheme data, to identify extinctions and colonizations, was based only on 'presence' or 'absence' at the individual sites. However, the synoptic nature of the scheme means that the index values can be used to provide additional evidence. This can perhaps be best illustrated by an example. The small skipper is one of the group of species which have expanded their ranges in recent years, and has colonized a few sites in the monitoring scheme close to the northern edge of its range. One such site is Gibraltar Point, a coastal nature reserve in Lincolnshire, at which the small skipper was recorded in 1979 after previous absence. There had been no records of this species since counts began here in 1974. Numbers increased rapidly (Figure 7.3), reached a peak in 1984 and have since fluctuated below this level. Without information from other sites, it would not be clear whether the rapid increase in index values, from zero to over 80 between 1978 and 1980 (Figure 7.3) was peculiar to Gibraltar Point or part of a wider increase in numbers. With the extra information from the

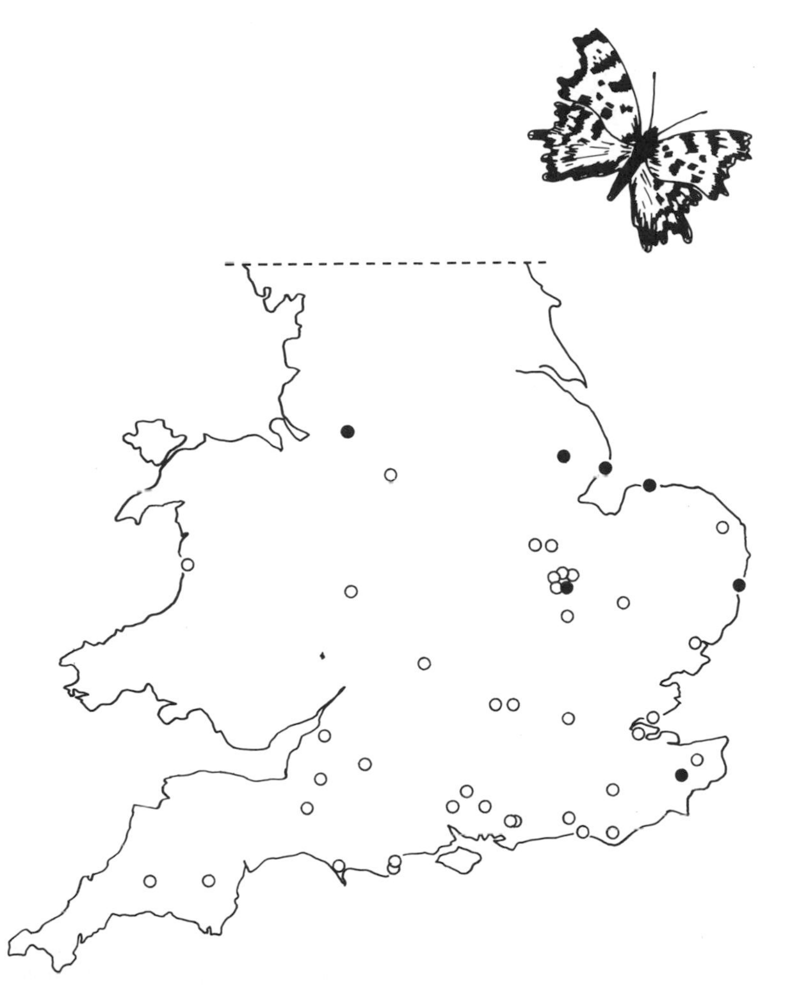

Figure 7.1 Expansion of range of the comma. Closed circles, sites identified as being colonized in the recording period; open circles, other sites with records of the comma in at least four consecutive years in 8 years of recording. Data to 1990. Redrawn from Pollard and Yates (1992). The comma has been spreading for many years and this spread is clearly continuing.

monitoring scheme, showing that general abundance remained very similar over these years, there can be no real doubt that the butterfly has recently colonized Gibraltar Point.

In addition to the confirmatory evidence of colonization, the patterns of increase and stabilization of populations promise to help with understanding the population ecology of butterflies. Do numbers build up to a carrying capacity and subsequently remain around that level, or do they increase and

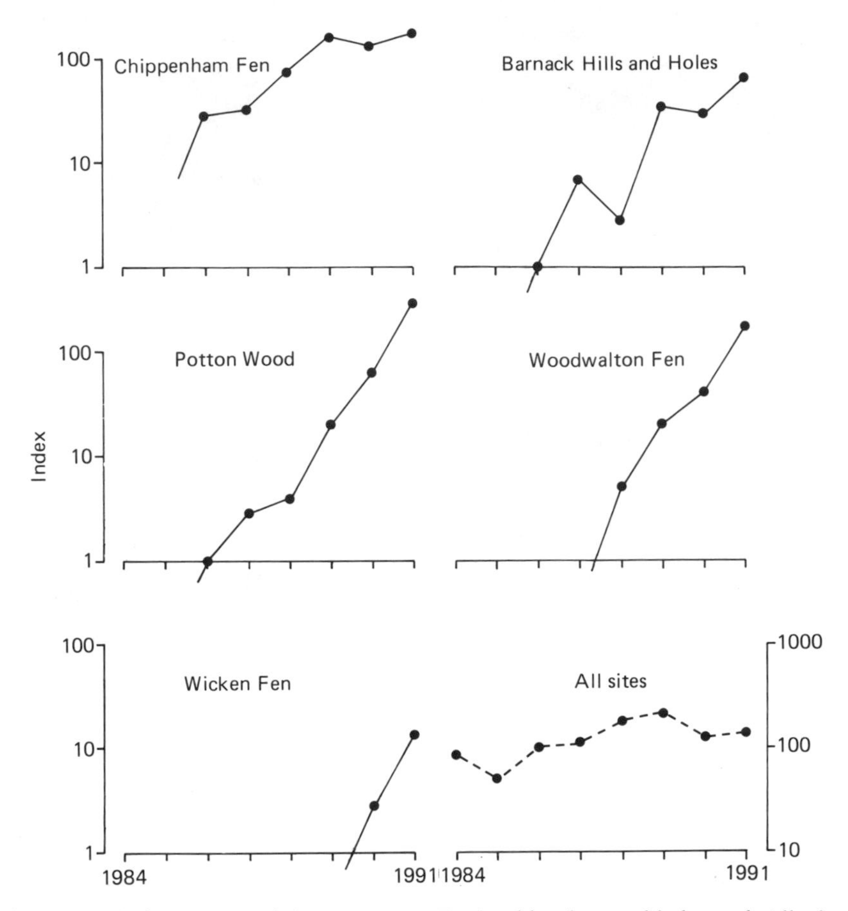

Figure 7.2 Colonization of sites in eastern England by the speckled wood. All of the sites were recorded from 1980 or earlier. At Potton Wood there were a few records earlier than those shown, although populations were, apparently, not established. At Chippenham Fen the speckled wood was recorded in low numbers in 1985, but there were insufficient counts for the calculation of an index of abundance.

then decline again, perhaps as their predators and parasitoids increase in numbers? So far, the examples available are very few and more are awaited with impatience!

The small skipper probably colonized Gibraltar Point in 1976 or 1977, although the first records were in 1978. The colonizing individual or individuals themselves are very unlikely to be seen and it may be a year or two before numbers build up to a level at which sightings are likely. For this reason, it is not possible to date colonization (or, for similar reasons, extinction) with precision.

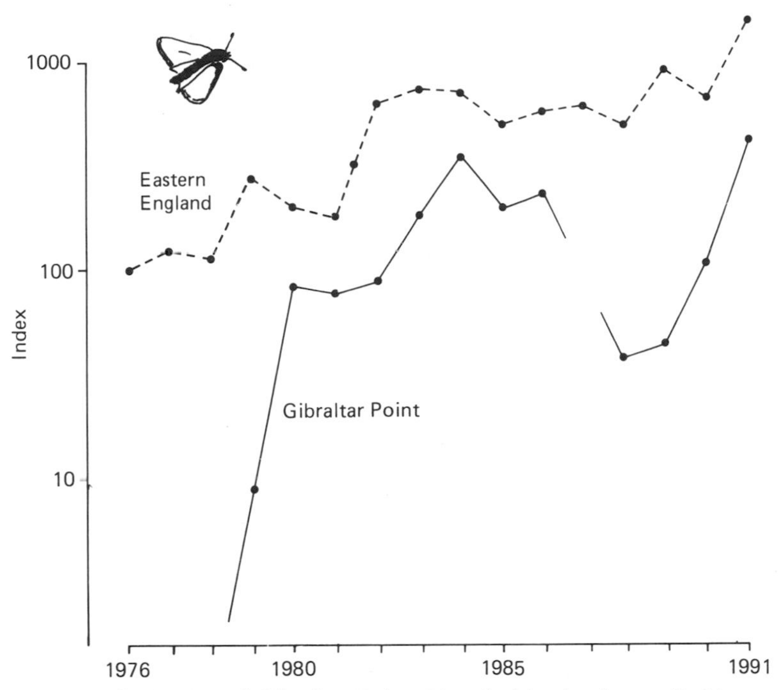

Figure 7.3 Colonization of Gibraltar Point, Lincolnshire, by the small skipper. There were no records from the start of recording in 1974 until 1979. There were insufficient counts for the calculation of an index in 1987. Following the initial rapid increase, trends at Gibraltar Point have been similar to those in eastern England.

It seems likely that dispersal and colonization are more frequent in hot, dry, summers. In such summers, not only are conditions good for flight, but individuals may also find local conditions inhospitable because of drought. Certainly, in hot summers, there have been records of butterflies, at monitored sites, away from their normal habitats. Examples include a single silver-studded blue on Ballard Down (Swanage) in 1983, single chalkhill blues in Ampfield Wood and Picket Wood in 1984 and one in Potton Wood in 1990; both of these species are considered to be sedentary, but these individuals had clearly dispersed from their breeding areas.

A variety of information suggests that several of our rarer butterflies have declined further during the period of monitoring from 1976 to 1990. In particular, Warren (1993) has documented losses of populations of many species in central southern England during the 1980s. As many of the monitored populations of rare species are on nature reserves, they might be considered reasonably secure. Unfortunately nature reserves are not immune from extinctions and there have been a few losses from monitored transects, such as the chalkhill blue at Kingley Vale and the high brown fritillary at Wyre Forest. For both of these species, there are sufficient comparative data from other sites for there to be little doubt that extinction has indeed occurred on these transect routes (although the chalkhill blue has survived elsewhere at Kingley Vale). As is discussed in Chapter 16, butterfly conservation is not an exact science; even if the requirements of a particular species are reasonably well known, the necessary management may be expensive or conflict with the needs of other plants or animals. In addition, even if ideal conditions are provided for a butterfly, there is always a chance that unfavourable weather or other adverse factors may eliminate a small population.

There have been a few examples of colonization of monitored sites by rare butterflies. The silver-spotted skipper, for example, has colonized the chalk downs of Old Winchester Hill; there is other evidence that this rare butterfly, which requires a short sward with some bare ground, has benefited from a recent upsurge in rabbit populations (Emmet and Heath, 1989) and the same is true of the Adonis blue (Thomas and Lewington, 1991).

The balance of losses and gains recorded in the monitoring scheme shows more colonizations than extinctions for both the common and rare butterflies, as listed in Chapter 1 (common butterflies, 103 colonizations, 65 extinctions; rare butterflies, 20 colonizations, 14 extinctions). However, the predominance of declines over increases in abundance of rare species in longer runs of data (Table 6.2) suggests that some populations of rare butterflies at sites in the scheme may be vulnerable. More generally (e.g. Thomas, 1991; Warren, 1992) there is little evidence that the decline of the rare butterflies has ended.

7.5 INTRODUCTIONS BY HUMANS

The release of rare butterflies, either into areas in which they had become extinct, or outside their known range, has long been practised. One famous instance was the introduction of the European map butterfly, *Araschnia levana* (L.), into the Forest of Dean in 1912 and its subsequent (reputed) extermination by an irate entomologist. The topic of introductions continues to cause controversy, although it is now generally accepted that they can make effective contributions to the conservation of some species (Oates and Warren, 1990).

From the narrow viewpoint of the academic scientist, interested in the natural response of butterflies to climate and changes in biotopes, introductions are unwanted complications. This is true even if the introductions are well documented, but many are anonymous and unrecorded. Most such 'unofficial' introductions, and many that are more carefully planned, are unsuccessful, but it is possible that the spread of a butterfly into a new area could stem from a single succeessful introduction. Similarly, there is little doubt that the colonizations identified in the monitoring scheme include a few deliberate, but unrecorded, introductions.

The known introductions at monitoring scheme sites were both eventually unsuccessful. The Adonis blue was introduced to Old Winchester Hill in 1981. The downland had been grazed by sheep, using small paddocks and rotation of grazing to create the short sward required by this butterfly. At first it flourished (Figure 7.4), but then populations declined and the last records were in 1988. An introduction of the wood white to Monks Wood in 1984 survived until 1988, but numbers were always low. Conditions for the wood white at the time of the introduction were thought to be good, but there is no doubt that they became less favourable over the following years, and the same is probably true of the Adonis blue at Old Winchester Hill.

Curiously, while the introduction of the Adonis blue at Old Winchester Hill failed, the silver-spotted skipper, which requires broadly similar conditions, colonized the reserve naturally and seems to have become established. Although the introductions to monitoring scheme sites were not successful, many other introductions survive to the present (Oates and Warren, 1990). However, while it is exciting to prepare a site for an introduction, considerable long-term commitment is required to maintain suitable conditions over the years and decades that follow.

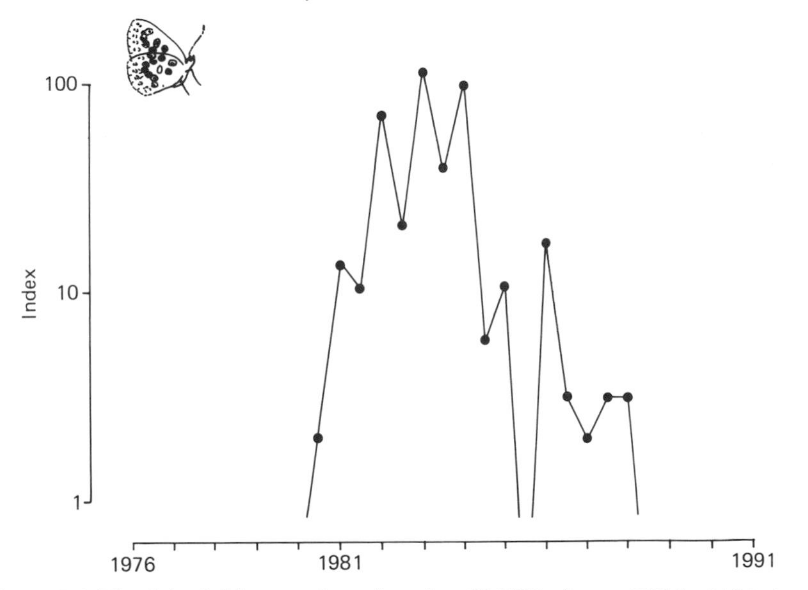

Figure 7.4 The Adonis blue was introduced to Old Winchester Hill in 1981; it flourished for a few years but then declined and is now, almost certainly, extinct again.

7.6 CONCLUSIONS

The monitoring scheme has been in progress just long enough for information on extinctions and colonizations to begin to emerge; in the longer term, these aspects promise to be amongst the most interesting in the scheme. The early results suggest (1) that species differ considerably in their patterns and frequencies of extinctions and colonizations within their range, while (2) the known changes of range of butterflies have been reflected, albeit to a limited and rather erratic extent, at sites in the scheme; in such cases, the pattern of establishment and build-up of populations may contribute to the understanding of the population dynamics of butterflies.

Considerable emphasis has been placed in this chapter on the difficulty in assessing whether extinctions and colonizations have occurred. Given the changes of range that are in progress, and those that are likely to occur in the event of climatic warming, the evidence may become overwhelming and make the present reservations academic.

– 8

Effects of weather on numbers

8.1 INTRODUCTION

Butterflies are associated in the mind with fine summer weather, partly because we are more likely to be out in the countryside to see them, but also because they fly only in warm weather. We might reasonably expect butterflies to become more abundant in warm conditions, and there is plenty of evidence to support such a view. However, the adult butterfly is only one stage in the life cycle; the abundance of the adults could depend on factors operating on the egg, larval or pupal stages; and these factors may be related to weather in different ways.

Many studies suggest relationships between weather and the abundance of butterflies. There is a limited amount of direct evidence from studies of the population dynamics of individual butterfly species and indirect evidence from apparent links between the ranges of butterflies and climate. There is also the evidence from the monitoring scheme that butterfly populations over large areas of the country fluctuate in synchrony (Chapter 6), and are clearly influenced by widespread factors, with weather the most likely candidate.

Some years ago, Beirne (1955) assessed the effects of various factors, including weather, on the abundance of butterflies and moths from 1864 to 1952. He had virtually no data, but based his study on anecdotal accounts in entomological journals. Using several thousand such references, Beirne assessed individual years as good, average or bad for Lepidoptera. For example, 1918 was 'good, especially for butterflies and spring larvae', 1919 'below average', 1920 'below average' and 1921 'average, but good for butterflies' and so on. From this unpromising material, Beirne drew the conclusions that the main factors in causing good years for Lepidoptera were (1) severe winter cold, thought to act by reducing the activity of predators and (2) warm dry summer weather; while the main factors causing bad years were (1) when the previous year was good, (2) high rainfall over several years, (3) droughts, and (4) late spring and summer frosts. It is of some interest to see whether our quantitative results, albeit over a much shorter period, support or contradict the conclusions from

Beirne's monumental qualitative study.

Recently, as global warming has become increasingly likely, the possible effects on butterflies have received considerable attention (e.g. Dennis and Shreeve, 1991). Will all resident butterflies benefit and those restricted to the south of England move northwards? Will new species colonize the country from continental Europe? Alternatively, will some of our northern butterflies become endangered? The many unknown factors in the responses to weather of butterflies and their food plants, means that predictions can be, at best, only in very general terms. The aim in this chapter is to identify some of the broad associations between weather and butterfly abundance, and, in so doing, help to predict the effects of climatic warming (Chapter 15).

8.2 WEATHER DURING THE MONITORING PERIOD

In a short period of recording, a study of the effects of weather on numbers may be inadequate simply because some conditions which would have major impacts on butterflies have not occurred. It is certain that our analyses are deficient for this reason. However, we have been fortunate; in the period 1976–90, several extremes of weather have occurred and the responses, or lack of responses, of butterfly species have been recorded.

The scheme began in a summer, part of which was described by the Meteorological Office (1977) as 'the longest hot spell for at least 100 years and probably much longer'. The recording period has included two winters, 1978–79 and 1981–82, with periods of severe cold, and four very hot and dry summers, 1976, 1983, 1989 and 1990. In general the summers of the early years, 1977–80, were cool, followed by warmer summers from 1981 to 1984 and then cooler again from 1985 to 1988. Although there have been exceptionally hot summers, there have been none in this period that have been exceptionally cold. Temperature data published by Manley (1974) suggest that there were several summers in the nineteenth century and the early years of this century that were colder than any experienced since then. There is of course no clear evidence in these weather data of a general warming; climatic warming, if and when it occurs, is likely to be detected only over several further decades.

8.3 ASSOCIATIONS BETWEEN WEATHER AND BUTTERFLY NUMBERS

8.3.1 Method of analysis

Our main approach to the study of effects of weather (Pollard and Lakhani, 1985; Pollard, 1988) has been by the use of a standard multiple regression model. The population index values in a given generation were assumed to depend on: (1) population size in the previous generation; (2) a particular weather variable; (3) unknown factors, including other aspects of weather; and (4) inaccuracy of recording and chance effects. The first two factors are included in the model equation and the last two considered together as 'error', i.e. the variability in the index values that is not accounted for by the previous index value and the particular weather variable considered.

The analysis was restricted to the commoner butterflies for which all-sites index values have been calculated (Chapter 6). For each butterfly species, up to 76 equations were calculated, using different weather variables, and significant associations were found (Pollard, 1988). The number of weather variables was kept to a minimum because of the small number of years of butterfly data available; even so, a large number of equations were examined and there is a strong probability that some very close associations were due to chance. A major difficulty is to distinguish between such chance results and those which are due to real effects of weather. We are confident that we have found some causal associations, but must expect that a few associations, at present believed to be causal, will prove to be due to chance.

Temperature and rainfall were the first obvious factors to consider. Both are likely to influence the survival of butterflies directly and indirectly, via effects on plant growth, disease, predation or other factors. There are many facets of both temperature and rainfall which could be considered, e.g. minimum winter temperature, spring frosts, long periods of extreme heat, exceptionally heavy rain, long periods of cool, damp, weather and so on. Of the many possibilities, just two extremely broad measures were included: monthly mean temperature and monthly rainfall. We included these weather data for the months of the previous year and those of the current year up to and including the month of the flight-period. The monthly weather variables were considered singly and also by pooling the data for 3-monthly periods. The results presented (Tables 8.1 and 8.2) are those using pooled, 3-monthly, weather data, as these produced clearer patterns of associations and, in particular, many more significant associations with rainfall.

For butterflies with two index values a year, changes from the second index in one year to the second index in the next year were examined. With these species the analyses can be taken further, to look separately at changes from the spring to summer generations, and from the summer to the following spring. This has already been done to a limited extent, for example to examine overwinter survival in the brimstone (Chapter 11).

The weather data used were the 'district' values for England and Wales (Meteorological Office, 1976–92). The collated butterfly index values include the Scottish sites but are dominated by those further south and the England and Wales weather data are probably the most appropriate. These district values are not precise measurements but are obtained by collation of weather data from many individual sites and so seem particularly suitable for this type of broad analysis.

8.3.2 Results

In the published analysis (Pollard, 1988), using the data for 1976–86, a larger number of significant associations between changes in butterfly numbers and weather variables were found than would be expected by chance. With a further 4 years' data, the number of significant associations has increased further and clearer patterns have emerged. There seems no doubt that there are strong effects of weather on butterfly numbers.

The strongest (most highly significant) relationships found were increases in index values when the weather immediately prior to and including the flight-period was warm and dry (Tables 8.1 and 8.2). Butterflies which showed this relationship most strongly included the small skipper, grayling, hedge brown, meadow brown, small copper, common blue, holly blue, wall, red admiral and small heath. This is perhaps the most predictable relationship and seems to confirm the 'natural expectation', mentioned at the beginning of the chapter, that butterfly numbers should increase in warm, dry, weather. For two species, the grizzled skipper and brimstone, increases in index values were significantly associated with warm and dry weather in the previous summer while, for the holly blue, warm weather in both previous and current summers seemed beneficial.

In eight species there were significant associations between increases in numbers and high rainfall in the early months of the previous year. This association was particularly pronounced in the ringlet and speckled wood; both species also tended to increase when the previous summer was cool.

There were few significant associations with temperature and rainfall in the winter before the current flight-period. In particular, there were no significant associations between cold winters and increase in numbers.

8.3.3 Interpretations

There is little doubt that the strong relationship between increases in butterfly numbers and warm, dry, summers is causal. Many British butterflies are scarce or absent in northern areas and it seems likely that the restriction to the south is, at least in part, related to summer temperatures (e.g. Dennis, 1977; Turner *et al.*, 1987). Virtually all of the British butterfly species are found where winter temperatures reach far lower levels than in Britain, and it is doubtful if any species are restricted in range here because of cold winters.

A life-table study of the white admiral (Pollard, 1979a; Chapter 12) suggests one possible mechanism for an effect of summer temperature on

Table 8.1 Weather and butterfly numbers. Significant ($P < 0.05$) positive and negative associations between temperature and butterfly index values (i.e. positive association = warm weather associated with increased index values). The temperature data used are the pooled means of three successive months; e.g. the 'June' associations are based on weather in May, June and July. The first and last months (January and August) weather data were thus used only in incorporation into the adjoining means. Table updated to 1990 from Pollard (1988)

	Previous year											*Current year*						
	F	M	A	M	J	J	A	S	O	N	D	J	F	M	A	M	J	J
Univoltine species																		
Small skipper																		+
Large skipper				–														
Dingy skipper																		
Grizzled skipper					+	+												
Brimstone					+	+				–								
Orange tip									+	+								
Green hairstreak																		
White admiral																		
Peacock																		
Dark green fritillary															–			
Silver-washed fritillary																		
Marbled white																		
Grayling			+															+
Hedge brown	+		+														+	+
Meadow brown			+														+	+
Ringlet				–	–	–												
Bivoltine species																		
Large white																		
Small white																		
Green-veined white					–	–												
Small copper																	+	+
Common blue			+														+	+
Holly blue	+	+	+					+					+				+	+
Wall			+			–							+				+	+
Indistinct voltinism																		
Red admiral																+	+	+
Painted lady																+		
Small tortoiseshell																+		+
Comma						–										+	+	
Speckled wood					–	–										+	+	
Small heath																	+	+

Table 8.2 Weather and butterfly numbers. Significant positive and negative associations between rainfall and butterfly index values (i.e. positive association = wet weather associated with increased index values). Pooled 3-monthly weather data, as in Table 8.1. Updated to 1990 from Pollard (1988)

	Previous year											*Current year*						
	F	M	A	M	J	J	A	S	O	N	D	J	F	M	A	M	J	J
Univoltine species																		
Small skipper										–								–
Large skipper			+	+										–				
Dingy skipper						–								–				
Grizzled skipper					–	–									–			
Brimstone						–												
Orange tip							–											
Green hairstreak			+											–				
White admiral													–	–				
Dark green fritillary						–								–				
Silver-washed fritillary																		–
Peacock													–					
Marbled white																	–	
Grayling																		–
Hedge brown																	–	–
Meadow brown																	–	–
Ringlet	+	+	+	+	+				–									
Bivoltine species																		
Large white										–								
Small white										–								
Green-veined white					+													
Small copper										–							–	–
Common blue																	–	–
Holly blue																	–	
Wall		+			+	+												–
Indistinct voltinism																		
Red admiral																		
Painted lady																		
Small tortoiseshell			+											–				
Comma	+												–					
Speckled wood		+		+	+	+												
Small heath																		

abundance. Although in the present analysis the white admiral showed no such significant effect, the strongest association with an individual monthly temperature was an increase in numbers when June was warm. In the study of the white admiral at Monks Wood, late instar larvae and pupae developed rapidly in seasons when temperatures were high; survival also was good. Predation by birds was shown to be important in these stages and it seems likely that rapid development, and consequent reduction in the time for which the larvae were available to birds, was the cause of the increased survival (Chapter 12). Similar considerations could apply more generally to those univoltine butterflies which appear to benefit from warmth: the small skipper, grayling, hedge brown and meadow brown. There is thus a possible link between the timing of the flight-period and butterfly numbers; as a generalization, when the flight-period is early and short there is a tendency for numbers to increase.

The other species in which increased numbers were associated with warmth are bivoltine or have a variable number of generations. In some of these cases, the benefits may derive in part from the production of additional generations. For example, in the case of the red admiral, the significant association with warm weather in April, May and June may be because a warm spring is favourable for early immigration, while more general summer warmth may permit an extra generation in the autumn. Similarly, the small tortoiseshell and comma have flexible life cycles; in warm summers, more individuals of these species develop quickly and breed a second time, whereas in cool summers a higher proportion enter hibernation soon after emergence from the pupae.

The bivoltine butterflies divide into two groups, the 'whites' which show little evidence of any response to warm and dry summer weather and the remainder, all of which seem to benefit considerably. In the cases of the small copper and wall, the inclusion of any third generation individuals in the second index of the year may contribute a little to the increased numbers. However, increase in abundance of the common and holly blues is equally associated with summer warmth and these butterflies have, at most, an insignificant third generation. The phenology of the common blue is discussed further in Chapter 11.

Most warm summers are also dry. For the present we have not attempted to separate out the possible effects of warmth and dryness. With data for more years, it will be reasonable to extend the analyses by including temperature and rainfall variables together in the models, and so assessing the separate contribution of each. There is limited evidence to suggest that some invertebrate predators are more active and effective in wet conditions (e.g. Dempster, 1967) and so, if this is true generally, temperature and rainfall may well have effects that are partly independent of each other.

The relationship between increased index values of the ringlet and speckled wood, and cool and wet weather in the previous year, is in accord with their preference for relatively cool and moist localities (although not with a general association between butterflies and warm weather). However, in the case of the ringlet the statistical association is to a large extent dependent on large fluctuations in the first 3 or 4 years of the scheme. Following the exceptional 1976 drought, index values of the ringlet fell dramatically in 1977 and then recovered during the generally cooler and wetter summers of 1977–80. Subsequent dry years seem to have had rather small effects on index values and if the analyses are repeated without the 1976 data, most of the significant results disappear. In the case of the speckled wood, the pattern of significant results remains much the same with and without the 1976 data and so the relationship with drought is more firmly established than for the ringlet.

The ringlet is univoltine and interpretation of an association with drought, if indeed the association is found to be causal, is fairly simple. Butterflies need moisture, either from dew, rainfall or nectar, while young larvae may suffer from the desiccation of their food plants (grasses). Thus a drought in one summer may have a direct effect on numbers of a susceptible butterfly in the following summer. In contrast, the complex succession of overlapping generations of the speckled wood makes interpretation more difficult; for example, the early summer generations may benefit from warm, dry weather because of residual moisture from the winter, while the later generations may be adversely affected by drought. Such an interpretation is consistent with the results of this analysis, as warm early summer weather in the current year seems to be beneficial (Table 8.1).

Several other butterflies, in addition to the ringlet and speckled wood, tended to increase following cool or wet summers. Butterflies showing this relationship include the green-veined white, which like the ringlet and speckled wood is often found in sheltered and humid areas, but also the wall and small heath which are usually found in relatively dry and open sites. The latter two species, like the speckled wood, also tended to increase when the current summer was warm.

There is evidence, in addition to that from the monitoring scheme, that several butterfly species suffered during the severe drought of 1976. For example, there was large mortality of white admiral larvae in Monks Wood (Pollard, 1979a) and little doubt that this was caused by the severe desiccation of the honeysuckle leaves. Thomas (1983b) also attributed the collapse of Adonis blue populations in 1976–77 to drought and noted that every leaf of its food plant, horseshoe vetch, died over extensive areas during the main period of summer feeding. Thomas concluded that many larvae must have starved to death. Such desiccation of food plants must vary considerably from site to site, depending particularly on soil characteristics; the shallow soils on the steep slopes of chalk downs, where the Adonis blue occurs, are likely to be particularly vulnerable to drought.

There can be no doubt that drought can have severe effects on butterflies, but it is likely that the relationships between rainfall and changes in butterfly numbers are not simple. If there is an optimum level of rainfall, below which drought causes problems and above which larvae suffer directly or via disease, then our simple linear models will be inadequate. In addition, high temperature and low rainfall interact to cause droughts; they are not solely caused by low rainfall. More complex models are clearly required for fuller descriptions of the associations between butterfly numbers and weather, but such models must await data for more generations of butterflies.

For two species, the grizzled skipper and brimstone, warm, dry weather was associated with increased abundance in the following year. Warren *et al.* (1986) showed a similar relationship for the wood white. In the case of the wood white, the key factors determining annual fluctuations were found to be egg-laying success and the survival of very young larvae, both of which were increased in warm, dry, weather. Similar considerations may apply to the grizzled skipper. Similar to the wood white, this species flies in the spring and early summer, when warm, dry weather may be particularly needed for egg-laying and good survival of young larvae; drought is unlikely to be a problem at this time of year. The dingy skipper, another spring butterfly, does not show this relationship with warmth, but numbers have increased when the previous summer was dry (Table 8.2).

The brimstone overwinters as an adult and it is possible that warm weather in the summer provides good conditions for the butterflies to feed at flowers before hibernation, leading to better winter survival. There is some evidence (discussed in Chapter 11) for other species which overwinter as adults, that an adequate food supply before hibernation is important. However, it is easy, with a little imagination, to suggest reasons for any particular association between abundance and weather; in some cases, as has been emphasized, the associations will be pure chance and are likely to disappear when a few more years of data are available.

In addition to the provision of information on effects of weather, the models used in the analysis also included previous population size. There are indications that several butterflies, which seemed to be badly affected by the 1976 drought, recovered in the succeeding cool summers because their populations were low, relative to the carrying capacity of the biotope, i.e. there was a density-dependent recovery. The inclusion of previous population size in the models enables such factors to be taken into account. As yet, this aspect has been little considered, but it may have an important role in population fluctuations.

8.4 TESTING THE MODELS

The main aim of the analysis so far has been to identify weather variables which have been associated with, and may have caused, changes in butterfly numbers during the period of monitoring. In the longer term, more complex models obtained in this type of analysis may be used to predict changes in numbers under particular weather conditions. It is already possible to test the available simple models by examining their success in prediction.

It was suggested in Chapter 6 that the hedge brown could be considered representative of a group of species with similar fluctuations; for this reason it is used as an example to test the predictive value of one of the simple models. The best-fitting model to the 1976–86 data (Pollard, 1988) included the 3-monthly, June–July–August temperature and the previous index value. The fit to the observed data was reasonably good (Figure 8.1). When this model, obtained from the 1976–86 data, was used to predict the 1987–90 results, it was encouraging to find that the general pattern of fluctuations was captured, although the predicted population level was much too high.

This modicum of success provides added evidence that summer temperature is indeed an important factor in the population ecology of the hedge brown (and probably of the other species with similar fluctuations), but also suggests, as expected, that summer temperature alone is inadequate for the prediction of its population fluctuations.

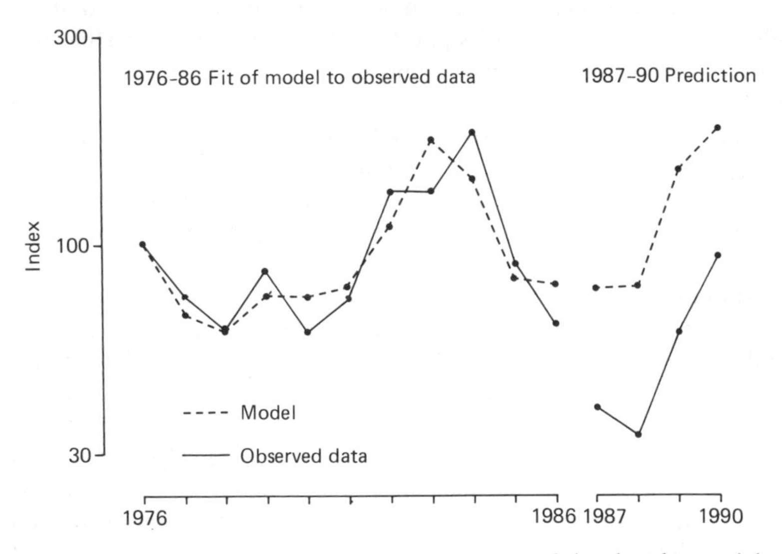

Figure 8.1 Association between summer temperature and the abundance of the hedge brown. A model was calculated from the 1976–86 'all-sites' index values and mean June, July and August temperature, and the fit to the recorded index values for this period is shown. The model was then used to predict the 1987–90 index values; the prediction was substantially wrong in 1987, but otherwise the model has performed well.

8.5 CONCLUSIONS

The analysis of associations between butterfly numbers and weather, presented in this chapter, is a first approach to a topic which should be developed much further as more data are acquired in the Butterfly Monitoring Scheme. A wider range of weather variables can then be examined and analyses of data from individual sites should also give a guide to geographical differences in responses to weather.

The results presented provide the first clear demonstration that butterfly numbers in general are influenced by weather and a start has been made in quantifying and interpreting these effects. We now return to Beirne's study, based on anecdotal information over some 90 years (Beirne, 1955), mentioned at the start of the chapter. The factors identified by Beirne as important included the main factors indentified in this study, i.e. beneficial effects of warm, dry, summers and harmful effects of droughts. However, our results do not suggest a beneficial effect of cold winters, identified by Beirne as the single most important factor, although there is a slight indication that dry winter weather may have been beneficial (Table 8.2). Nevertheless, it is surprising how much common ground the two approaches have. It may be questioned, however, whether it is very useful to consider years as 'good' for Lepidoptera as a whole. Species may vary greatly in their patterns of fluctuations and in their responses to weather.

The effects of late spring frosts, also mentioned as important by Beirne, have not been examined in the monitoring data. More generally, it is implicit in these analyses that it is the broad pattern of variation in weather from year to year that largely determines the fluctuations of butterflies. Ehrlich *et al.*'s (1972) observations on the catastrophic effect of a summer snowfall in Colorado, USA, is a reminder that such very occasional, but extreme, conditions may have impacts which last for years. So far, only the 1976 drought seems to have had a dramatic effect on monitored populations in Britain.

Weather does not only affect fluctuations in abundance. It also has a major role in the timing of the flight-periods of butterflies and in the number of generations in a season, as will be discussed in Chapter 10. Almost certainly, weather plays a role in dispersal and the colonization of new sites. It may also influence the range of biotopes in which a butterfly can breed. Thus the effects of weather on abundance, discussed in this chapter, have to be viewed in the context of effects on most, if not all, aspects of butterfly biology. A synthesis of these effects of weather is attempted in Chapter 15, where the likely impact of climatic warming is discussed.

– 9

Migration

I watched from on board ship a steady southward movement of butterflies [painted ladies], fifty miles or so out to sea off the west coast of Africa, from the mouth of the Mediterranean to as far south as Sierra Leone. These had undoubtedly come south from Europe . . . .

C.B. Williams, *Insect Migration*, 1958.

9.1 INTRODUCTION

The migration of butterflies and other insects, over hundreds or even thousands of kilometres, is quite remarkable. The imagination is captured by the thought that such apparently fragile creatures can cross continents and sometimes oceans. In particular, the travels of the monarch butterfly, *Danaus plexippus*, in America have greatly interested and excited ecologists. The demonstration that some Canadian monarchs overwintered in spectacular aggregations in Mexico was the culmination of many years of study of these butterflies (Urquhart, 1976; Urquhart and Urquhart, 1978).

Migration in butterflies, or other insects, is not strictly analogous to migration in birds. In the case of butterflies, individuals do not, except perhaps by chance, return to the areas in which they were born, whereas birds may do so year after year. Monarch butterflies, for example, fly south in the autumn to overwinter and then most breed in the spring fairly close to their overwintering sites; it is mainly adults from subsequent generations that reach the far north again.

The definition of migration is not simple and the term tends to be used in different ways for different groups of organisms. Taylor (1986) includes four types of movement by animals in his definitions of migration, including redistribution of individuals between generations. Only one of the four types is the two-way, actively controlled, seasonal movement commonly regarded by naturalists as migration. Solely for the purposes of this book, a definition based on geographical boundaries rather than on the behaviour

of the insects is adopted, i.e. migrant butterflies are considered to be those in which numbers in Britain depend (or are believed to depend) to a large extent on immigration from abroad.

Butterfly migration is inherently difficult to study. In the case of birds, much has been learnt from ringed individuals, but few studies of this type have been made of marked butterflies, as the chances of recapturing individuals is extremely small. The classic early work of Williams (1930), in collating observations from many countries, demonstrated that large-scale movement of butterflies is a common phenomenon in many parts of the world. In Britain, many thousands of records of migrant butterflies and moths have been collected and published annually, first by T. Dannreuther, starting in 1931, followed by R.A. French and later by R.F. Bretherton and J.M. Chalmers-Hunt, giving an invaluable record of the location of immigrations and some indication of the relative abundance of migrants from year to year. Compared with these 60 years of records, the data from the monitoring scheme are based on few sites over few years, but they have the advantage of systematic recording and the data make a distinctive contribution to the study of migrant butterflies.

In Chapter 11, monitoring data which suggest movement of the brimstone between hibernation and breeding areas are described. Such movements could be regarded as incipient migration. In this chapter, different types of migratory behaviour in British butterflies are discussed. Then the characteristics of fluctuations in abundance of the large and small white butterflies are examined to discover whether they show the erratic changes expected of migrants. Finally, the commonest of the major migrants, the clouded yellow, red admiral and painted lady, are considered in more detail.

9.2 MIGRATORY BEHAVIOUR IN BRITISH BUTTERFLIES

Ford (1945) quotes records of the wall, small copper, meadow brown and other species that are usually considered to be sedentary, at Light Vessels some kilometres off the coast of Britain. Presumably, these butterflies had been displaced by unusual weather conditions, but nevertheless their occurrence out at sea suggests that individuals of some, basically sedentary, species may, on rare occasions, reach Britain from the continent. However, these chance arrivals are not here considered to constitute migration.

Another group of butterflies is known to range widely over large areas of countryside and some of these have been seen in large numbers, apparently on migration, at sea or on the coast. This group includes the peacock and small tortoiseshell, but paramount amongst them are the large and small white butterflies. It has been suggested that mass immigration from the continent plays a major part in determining the numbers of these crop pest species that are present in this country each year (e.g. Frohawk, 1934; Ford, 1945).

The most unequivocal migrants are those butterflies which breed in Britain but overwinter here so rarely that the contribution to the following year's population is negligible. The three most common of these migratory butterflies in Britain are the red admiral, painted lady and clouded yellow, in declining order of average frequency. Although individual red admirals have been known to hibernate in Britain (e.g. Frohawk, 1934), there is little doubt that this is a rare occurrence. The painted lady and clouded yellow breed continuously, with no adaptation for diapause; overwinter survival is thus possible, if indeed it ever occurs, only in exceptionally mild winters.

Baker (1972b) considered that the movements of individual red admirals and painted ladies are mainly short range, with flight from the near continent, as opposed to long range, from the Mediterranean region or

North Africa. He has shown, from measurement of museum specimens, that British individuals of the red admiral and painted lady generally have a larger wingspan than those caught in the Mediterranean. Baker concluded that the disparity in size suggests that the British and near continental stock is genetically different from that originating in southern Europe. An alternative possibility is that large individuals are capable of longer flights and so, the farther north a sample is taken, the larger the butterflies are likely to be. The patterns of occurrence of these species at sites in the monitoring scheme will be considered in the light of Baker's ideas.

9.3 THE LARGE WHITE AND SMALL WHITE

The large and small white butterflies are pests of cultivated brassicas and there is no doubt at all that they fly long distances. There are few patches of cabbages or brussels sprouts, however small and remote, which escape their attention. Wild food plants are also used, but it is likely that these are of minor importance in Britain. Migration in these whites has been viewed in two ways. Firstly, attention has been focused on occasional mass migrations such as those detailed by Williams (1930). More generally, Williams and others clearly considered that these migrations are intermittent events. On the other hand, Baker (1968, 1978, 1984) suggested that each individual butterfly of these species is more or less permanently migrating, stopping only intermittently to feed at flowers, mate and lay eggs; thus migration is not seen as intermittent and characterized by mass flights, but may involve steady movements at low density and is considered by Baker to be an intrinsic part of the life of these butterflies.

If intermittent migrations occur and have a major impact on numbers of these butterflies in Britain, one would expect this to be reflected in both the nature of the population fluctuations from year to year and in the regularity of the counts of butterflies over each recording season; index values should fluctuate considerably, according to whether or not large migrations occur in a particular year. However, in the cases of both the large and small whites, the year-to-year fluctuations in index values are no more variable than in other species that also have two, or sometimes three generations each year (Chapter 6). In addition, in the case of the large white, there is a strong relationship between the index values of the first and second generations each year,* i.e., in Britain, numbers in the second generation are quite closely related to numbers in the first generation of the same year. This is evidence against a major effect of large, intermittent migrations.

The pattern of weekly counts of the large white each season reinforces the impression of regularity and of a continuity of populations (Figure 9.1); the

* All-sites data for the large white, 1976–91, correlation coefficient $r = 0.59$, $P < 0.05$.

rise and fall of counts in each generation is fairly smooth, suggesting gradual emergence (perhaps of local populations in the farmland of eastern England) or alternatively the steady arrival and departure of migrants in the way suggested by Baker (1978).

The monitoring data for the small white are rather different and have an unusual feature. The second generation index (which includes any third generation individuals) has been much larger than that of the first generation in every year, and the difference is more pronounced than in any other species. The reason for this large difference is not known, but it may be associated with migration, i.e. the second generation may be augmented by immigrants. Unlike the large white, the small white shows no significant correlation between the first and second generation indexes.* This result also suggests that numbers in the second generation may be influenced by the number of immigrant individuals rather than by the number of individuals breeding in the first generation in Britain. If such immigration does indeed occur, it appears to be as a steady flow of individuals rather than erratic influxes, as shown by general regularity in the pattern of counts at sites in the scheme. A typical example is that for a chalk down site in southern England (Figure 9.2). A downland site has been chosen for illustration, because at these dry, open sites a similar butterfly, the green-veined white, is relatively rare and misidentifications are unlikely to have much effect on the results. The regularity in the pattern of counts shown at this downland site is typical of the small white counts more generally; only very rarely have there been sudden, large increases which might possibly be caused by local concentrations of migrants.

In summary, the monitoring scheme data provide little evidence that intermittent, large migrations have played a major role in the population fluctuations of the large and small whites in the recording period, although monitoring evidence alone cannot be conclusive. The data for the small white in particular are consistent with low-density migration of the type suggested by Baker (1978).

It is likely that both of these butterflies are now less abundant than they were in earlier decades. Feltwell (1982) suggested that numbers of the large white have been reduced by granulosis virus, introduced by immigrant butterflies in 1955, although the evidence for an impact of the virus on numbers is not given. Another possibility is that the development of efficient insecticides, in the 1950s, has depressed numbers of both species, although, again (because there was no monitoring at the time) evidence is lacking. Thus, in earlier years, very large numbers of butterflies may, on occasion, have emerged from brassica crops in eastern England and perhaps on the near continent, over a short period. Similar emergences, combined with

* All-sites data for the small white, 1976–91, $r = 0.31$.

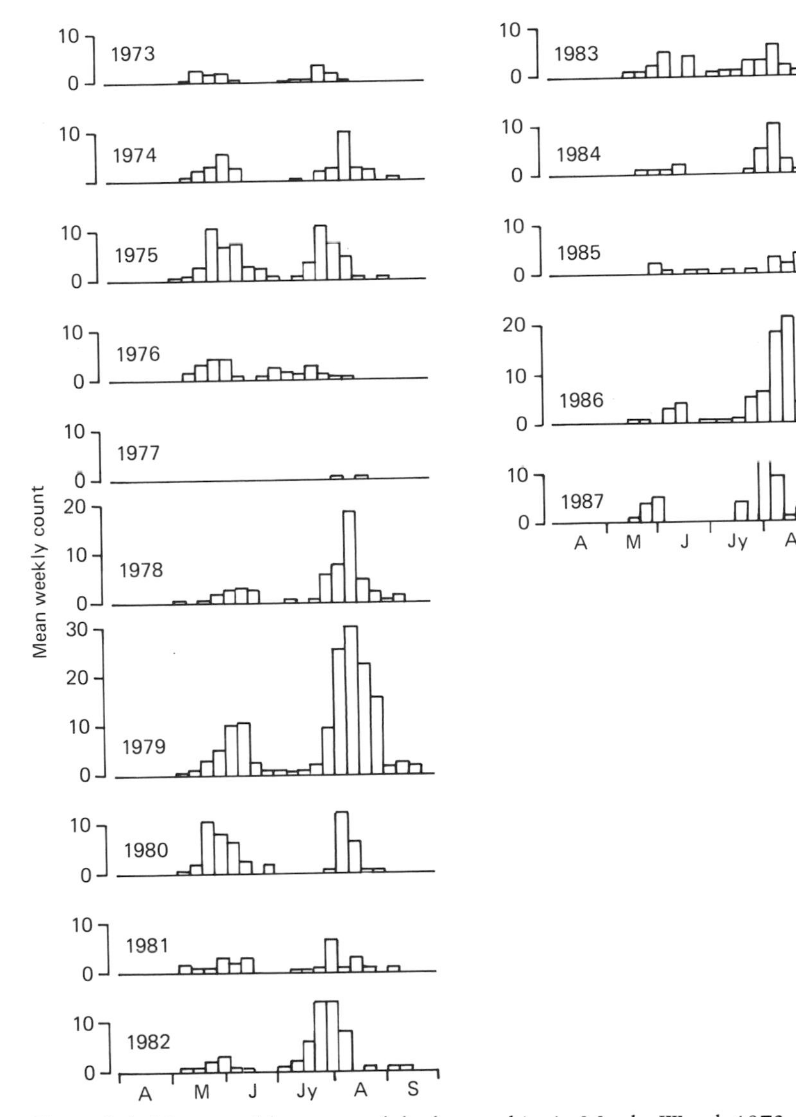

Figure 9.1 Mean weekly counts of the large white in Monks Wood, 1973–87, the period in which several counts were made in the wood in most recording weeks. The consistency in the timing of the two generations and the regularity in the pattern of counts suggests the absence of large and erratic immigrations in this period.

particular weather systems, could have led to the dense concentrations of these butterflies which have been documented so graphically. Conspicuous migrations at the northern edge of the range (e.g. Vepsalainen, 1968) may

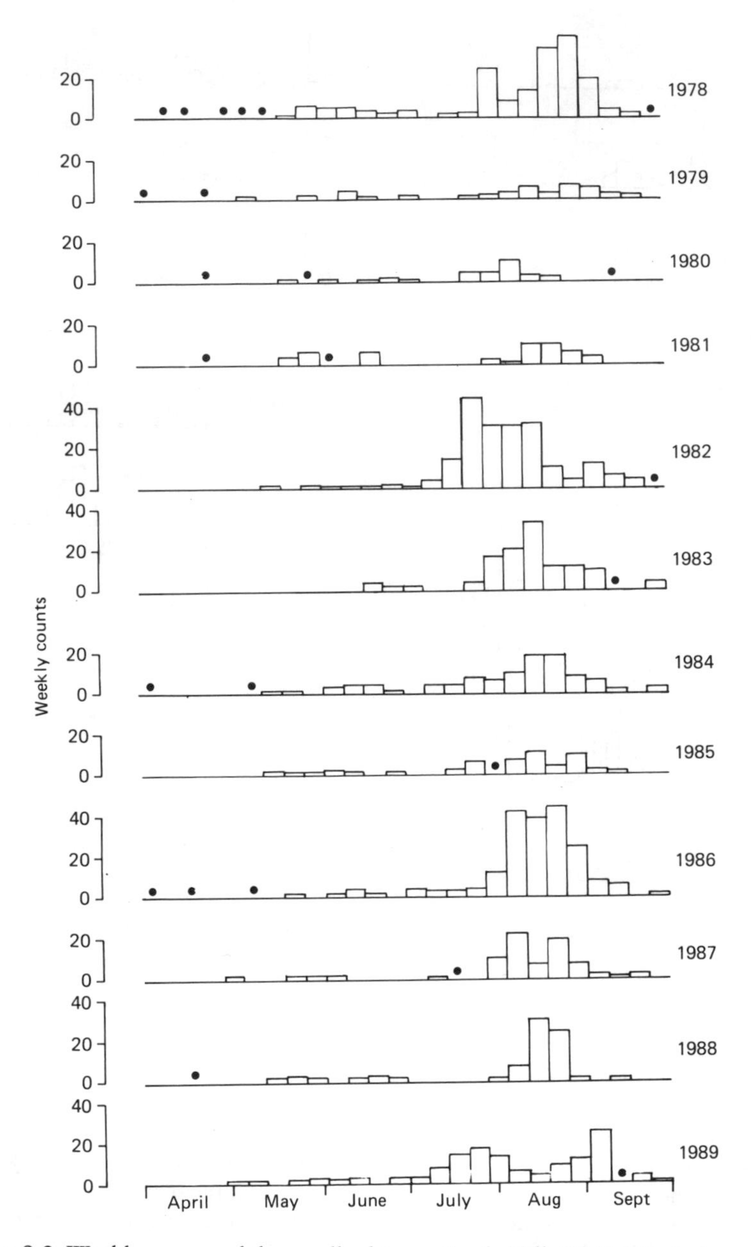

Figure 9.2 Weekly counts of the small white at Castle Hill, 1978–89. A dot indicates no count in that week. The spring generation is consistently small, compared with the summer generation, and it is possible that the summer generation is augmented by immigrants. There is a clear third generation only in 1989.

also be associated with years of exceptional abundance.

9.4 THE MAJOR MIGRANTS

The three, relatively common, major migrants to Britain are the clouded yellow, red admiral and painted lady. There are several other rarer migrants, including Berger's clouded yellow, the pale clouded yellow, the Camberwell beauty and the Queen of Spain fritillary, but as yet we have little or no data from the monitoring scheme on these species.

9.4.1 The clouded yellow

The clouded yellow is resident at low levels in North Africa, the Mediterranean area and elsewhere in southern Europe (Bretherton and Emmet, in Emmet and Heath, 1989) and migrates, in spring and summer, to much of northern Europe. The larvae feed on clovers and other legumes, usually in open situations.

The clouded yellow has been abundant in only one year, 1983*, in our recording period (Table 9.1). There is some indication from these data that 1983 was one of a period of several years in which numbers were above average in Britain. Semi-quantitative information (Williams, 1958) suggests that the clouded yellow has had similar periods of scarcity and abundance for over 100 years, perhaps reflecting the general abundance of the species in southern Europe. Before 1983, the previous year of comparable abundance was 1947 (Bretherton and Emmet, in Emmet and Heath, 1989) and invasion years now seem rarer than earlier in the century. However, there were some long gaps between 'clouded yellow years' in the nineteenth century.

Bretherton and Emmet mapped the distribution of sightings of the clouded yellow in 1983, as recorded by entomologists from all over Britain. Their map suggests that peak numbers were in the southwest of England.

Table 9.1 Number of sites at which the clouded yellow was recorded, and the sum of index values, in each year. Updated (and corrected) from Pollard *et al.* 1984

	1976	*77*	*78*	*79*	*80*	*81*	*82*	*83*	*84*	*85*	*86*	*87*	*88*	*89*	*90*	*91*
No. of sites	1	2	0	2	4	6	5	54	9	1	1	2	0	6	11	3
Sum of index values	1	3	0	2	21	16	19	653	33	1	2	3	0	28	40	5

* 1992 was another year of abundance, second only to 1983

The first sightings on counts in the monitoring scheme were at two sites in Dorset and, surprisingly, at the Peak District site in Lathkill Dale in Derbyshire, all on 8 June. During the course of the summer, there were records from nearly all of the sites in the scheme except those in the north of England and Scotland, although we know that some clouded yellows were seen in Scotland.

The seasonal pattern of counts of the clouded yellow in 1983 (Figure 9.3) suggests that there was an initial immigration in early June and that these individuals bred in Britain and produced a second generation which peaked in August. A third generation seems possible, but if so this fell only partly within the recording period. This view was not fully shared by Bretherton and Chalmers-Hunt (1985), who considered that further immigrations largely accounted for the great abundance in July and August. Either interpretation is possible, but the smooth rise and fall of the summer counts (Figure 9.3) is more consistent with local breeding. Certainly, there were many observations of egg-laying and of larvae (e.g. Thomas and Lewington, 1991).

Although the clouded yellow is a migrant and spends much of its time in purposeful flight across country, there were nevertheless some areas where the butterflies congregated and where large numbers were recorded. Many individuals were seen at coastal sites, in clover fields and in flower-rich

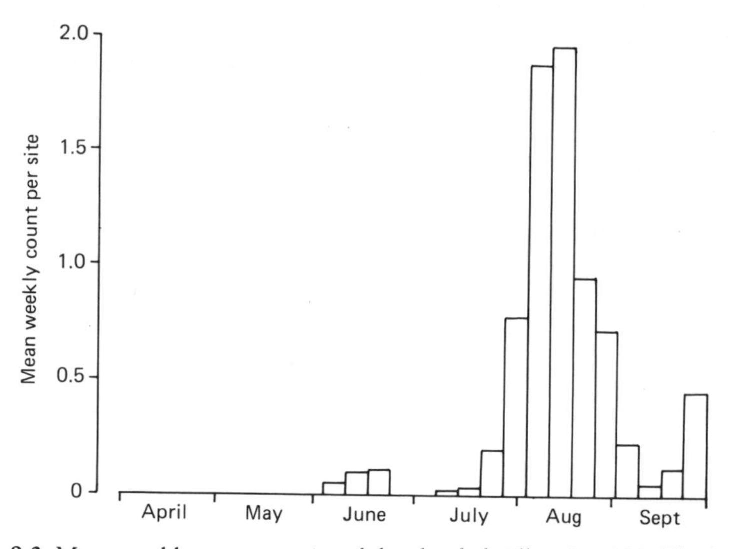

Figure 9.3 Mean weekly count per site of the clouded yellow in 1983. The June butterflies are immigrants, the peak in July and August may be the result of breeding within Britain or may include further immigrants. There is a strong indication of a third generation at the end of the recording period. Redrawn from Pollard *et al.* (1984).

grasslands. At Ballard Down, a chalk grassland site near the coast at Swanage in Dorset, a total of 112 were seen on transect counts through the season.

It is clear, from this account, that Britain is very much peripheral to the area typically occupied by the clouded yellow and the causes of its population fluctuations must be sought elsewhere. A sharp decline in abundance in Britain of a migrant bird, the whitethroat, *Sylvia communis*, in the 1960s was thought to be related to drought in the Sahel zone of West Africa (Winstanley *et al.*, 1974). In a similar way, it is possible that in monitoring numbers of butterfly migrants in Britain, we are gaining some information on the changing suitability of conditions where core populations occur. Unfortunately, the sources of the British immigrants are at present poorly known.

9.4.2 The red admiral

The second of the major migrants to Britain, the red admiral, provides a sharp contrast with the clouded yellow. The red admiral is much more predictable in that it arrives here in large, if variable, numbers each year and appears to breed here successfully every year. In some, perhaps most, years some adults survive the winter. The red admiral reaches high latitudes, with records from close to the arctic circle (Emmet and Heath, 1989), and in Britain larvae have been found as far north as the Orkneys (Lorimer, 1983). In the monitoring scheme the butterfly has been recorded in moderate abundance, in some years, at one of the most northerly sites, Sands of Forvie in Aberdeenshire. Some red admirals are usually flying in October, after the recording season has finished, and for this reason our information is incomplete. The food plants are predominantly nettles, often in sheltered woodland or hedgerow situations.

The pattern of migration is not well understood. Bretherton and Emmet (in Emmet and Heath, 1989) suggest that very early individuals, in April, may fly from North Africa, while later immigrants come from progressively further north. This contrasts sharply with Baker's view, mentioned earlier in this chapter, that most red admirals are short-range migrants from the near continent. If Baker's view is correct, April individuals would be more likely to have overwintered in Britain or the near continent, rather than have flown much farther than the later arrivals.

Clearly there is much to be learnt about the migration of this common butterfly. What can the monitoring data contribute? Firstly, the annual fluctuations of the all-sites index values show that, compared with the other common migrant, the painted lady, numbers of the red admiral are relatively stable (average yearly change × 2.0, compared with × 6.0 for the painted lady; see Chapter 6). This stability suggests some regularity in the pattern of migration of the red admiral. On the assumption that greater distance of travel will result in greater variability, such regularity is more consistent with short-range movement from the near continent rather than from more distant regions.

We can also examine the seasonal occurrence of the red admiral at the monitored sites. Again, compared with the painted lady (Figures 9.4 and 9.6), the seasonal pattern is very regular. Whereas the painted lady has sudden peaks of abundance, often following periods with no records at all, the weekly counts of the red admiral rise and fall in a manner suggesting successive generations arising from more or less local breeding.

In summary, our data lend support to Baker's (1972b) view that migration of the red admiral to Britain is, predominantly, from relatively

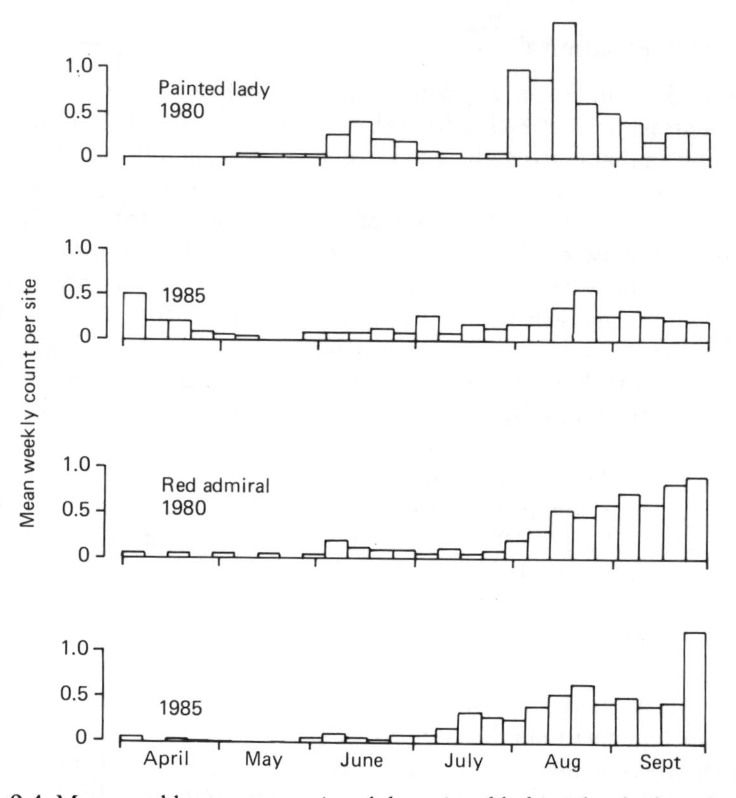

Figure 9.4 Mean weekly counts per site of the painted lady and red admiral in 1980 and 1985. In these and other years (see also Fig. 9.6) the seasonal patterns of the painted lady counts are very different, while those of the red admiral are similar from year to year. The contrast between the species suggests different migratory behaviour. Redrawn from Pollard *et al.* (1986).

close sources. Certainly, the results contrast very sharply with those of the painted lady described below.

9.4.3 The painted lady

The painted lady is very erratic in its appearance and abundance in Britain. Its population fluctuations are easily the most unstable of all of the commoner butterflies (Chapter 6). Other butterflies which show large fluctuations in numbers, such as the holly blue and wall, undergo periods of abundance and scarcity lasting several years, but in the case of the painted lady, numbers fluctuate enormously from one year to the next.

This butterfly is widely distributed in most parts of the world. Its migratory behaviour has been studied in North America (e.g. Abbot, 1951; Tilden, 1962) and in the Mediterranean (e.g. Larsen, 1976). In both these areas, its migrations were thought to be irregular, with no indications of return flight; however, in this country, Baker (1972b) considered its migrations to be short range, similar to those of the red admiral, moving first northward and, later in the summer, with a southward return flight.

The seasonal pattern of counts differed greatly from year to year. In the examples illustrated (Figures 9.4 and 9.6), the red admiral was fairly consistent in its seasonal pattern, but the painted lady was sharply different in each year. In 1980, the year of greatest abundance, there were large migrations of the painted lady early in June and at the end of July; in 1985 it was recorded quite widely in the first recording week at the beginning of April, while in 1990 there were more records in early June than at any other time.

The 1980 migrations of the painted lady were certainly the most spectacular in the recording period; they have been described in some detail by Bretherton and Chalmers-Hunt (1981) and by Pollard (1982a), the latter based on the monitoring scheme results. The early June records (Figure 9.4) were from the west of Britain, especially the west coast. The wind direction at this time was southwesterly. It seems possible that these individuals arrived from North Africa, first flying northwestwards out into the Atlantic before turning back towards Britain around a high-pressure weather system. At the end of July there was a further spectacular immigration, this time on the east coast from East Anglia northwards. Counts at monitoring scheme sites show this immigration clearly (Figure 9.5). The majority of arrivals seem to have been on 29 and 30 July and the wind direction on these days was southeasterly around a high-pressure area centred to the north of Scotland. Bretherton and Emmet, in Emmet and Heath (1989), noted that this migration was seen as far away as Finland, and this gives an indication of the enormous scale of movements that may be involved.

Bretherton and Chalmers-Hunt (1981) concluded that there was little

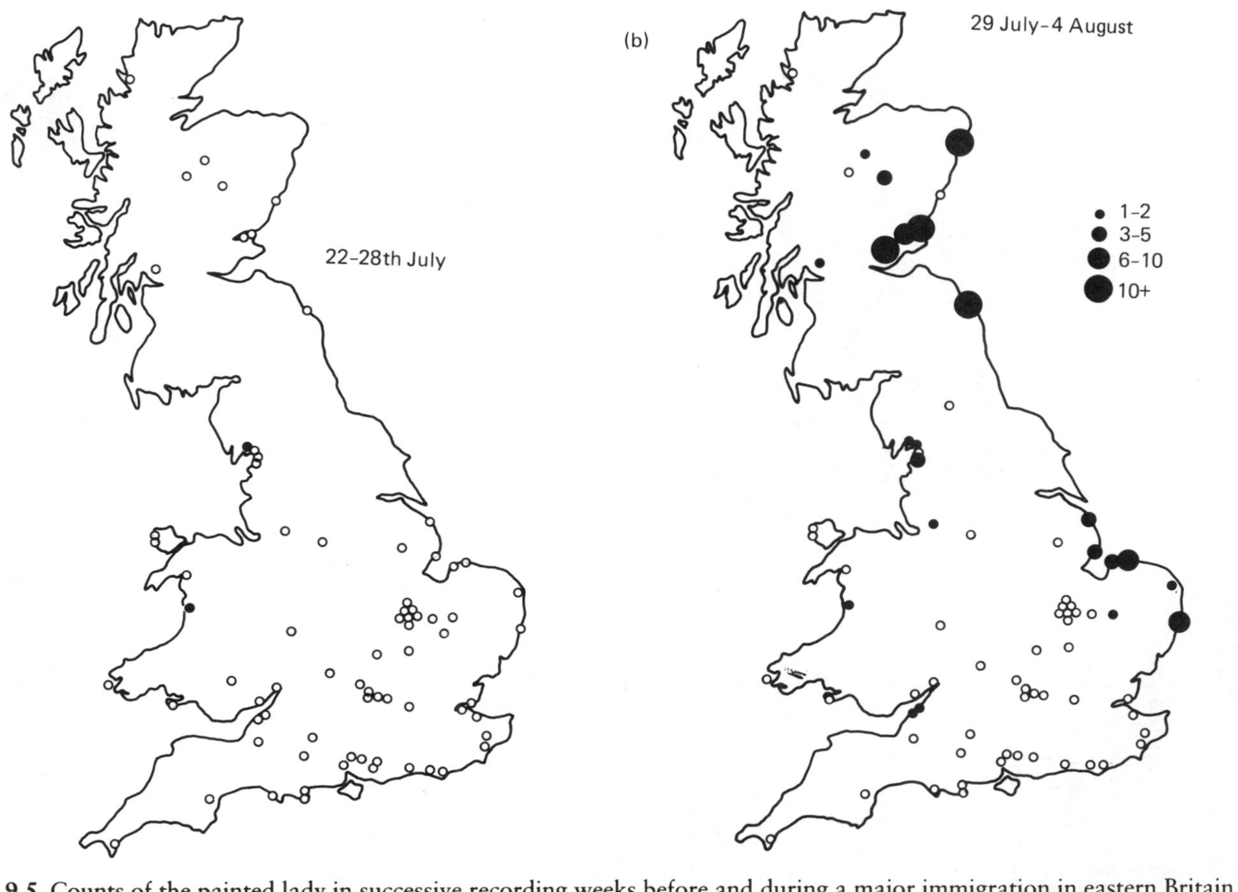

Figure 9.5 Counts of the painted lady in successive recording weeks before and during a major immigration in eastern Britain in 1980. Open circles indicate that a count was made, but no painted ladies were recorded. Redrawn from Pollard (1982a).

successful breeding by the painted lady in this country after either migration. Certainly, individuals making up the second peak, in late July, could not have been the progeny of those in early June; the time interval was too short, especially as the summer of 1980 was cool. Larvae from eggs laid by the second wave of immigrants must certainly have died when they were overtaken by the cool weather of autumn (e.g. Winter, 1981).

The 1980 migrations of the painted lady were exceptional in their abundance and perhaps in the short time span of the major movements; however, in other years similar erratic movements have occurred. It is difficult to reconcile this pattern of migrations with the regular and systematic short-range movements, first to the north in the early months of the year and later to the south, envisaged by Baker (1972b). Results from the monitoring scheme present a quite different picture, that, as in the case of the clouded yellow, Britain is very much on the periphery of the vast area within which the painted lady moves in an erratic manner and that the patterns of movement are, to a great extent, influenced by weather systems.

It seems likely that the migrations of the red admiral and painted lady are very different in their nature. The results from the monitoring scheme suggest that while the red admiral may be, predominantly, a short-range migrant, the painted lady is truly a traveller across entire continents.

9.5 THE BEGINNINGS OF AN INTERNATIONAL VIEW

As migrations may involve flight across many countries, a much fuller picture would emerge if information were available from Europe and beyond. Williams (1958), for example, quoted records from The Netherlands on the numbers of sightings per year of the common migrants and showed a close correlation with changing abundance in Britain. In 1989 a monitoring scheme, equivalent to the British Butterfly Monitoring Scheme, began in The Netherlands. In 1990, there were sufficient data from the Dutch scheme to enable weekly counts to be plotted (C.J. van Swaay, unpublished data).

The weekly counts of the red admiral and painted lady in Britain and The Netherlands (Figure 9.6) are presented to show the first of these extended data on migrants. The early June migration of the painted lady in England was not noticeable in The Netherlands, although the flight-periods were very similar, while the general pattern of counts of the red admiral in the two countries was also similar. Results from just a single year are unlikely to be very enlightening, but when comparisons can be made over several years, and perhaps extended to other countries, the understanding of these migrations is likely to be advanced considerably.

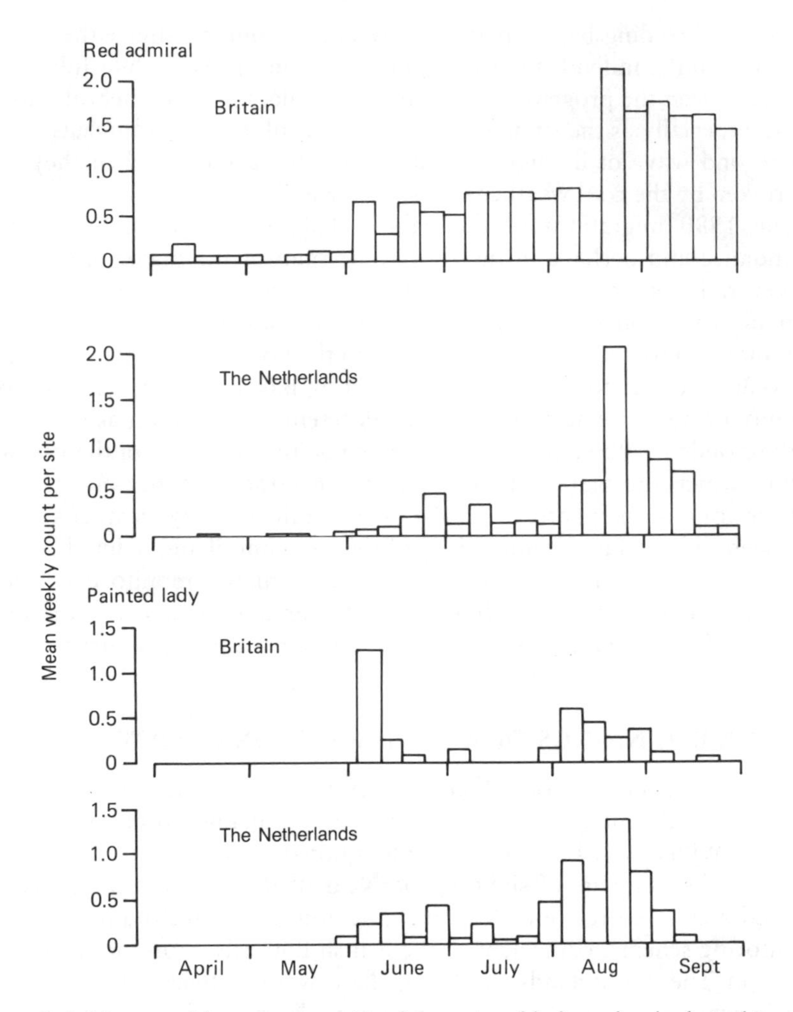

Figure 9.6 Mean weekly counts per site of the painted lady and red admiral in Britain and The Netherlands in 1990. The June immigration of the painted lady in Britain, shown by the high counts following a period with no counts, is not clearly apparent in The Netherlands. No clear conclusions emerge from this comparison, as might be expected from results from a single year; however, over a period of years counts of migrant butterflies in different countries will undoubtedly add to an understanding of their migrations.

9.6 CONCLUSIONS

Some of the disagreements by entomologists on the nature of butterfly migrations may stem from the assumption that the migratory process is basically the same in all species. Walker and Riordan (1981) and Walker

(1991) have shown that migration in a group of butterfly species in Florida was steady and directed over periods of many days, and in several of these species, flights were first to the north and later to the south. These studies were made using remarkable traps, which captured migrating butterflies over a 6 m front. Walker's results, although with different species, support those of Baker (1984) for the small white and there can be no doubt that such directional flights occur in many species.

On the basis of their results, Walker and Riordan (1981) questioned the importance of synoptic-scale wind systems for butterflies more generally. We believe that there is sufficient circumstantial evidence, from the monitoring scheme and from other studies, that major migrations of the painted lady are associated with wind systems. There is certainly no doubt that the mass movements of some other insects, such as the African armyworm moth, *Spodoptera exempta* (Rose *et al.*, 1985), are associated with weather systems and often influenced by winds. As migration is likely to have evolved in response to a variety of factors, there is no reason to assume a common mechanism in all species. Thus, in Britain, the large white, small white and red admiral may fly first northwards and later south within this country and to and from the near continent; in contrast, the painted lady and perhaps the clouded yellow may cross continents and arrive where a weather system determines.

While the evidence from butterfly monitoring cannot resolve questions on the nature of butterfly migration, there is no doubt that it can make some contribution. As yet, the full potential of the monitoring data for the study of migrants has not been realized. For example, we have yet to examine the geographic changes in the distribution of migrants during the course of each year, except in the case of a few spectacular immigrations. The migrants could, because of their mobility, be among the first species to respond to climatic change. Whether or not this proves to be the case, there is no doubt they will continue to provide fascinating subjects for study.

10

The flight-periods of butterflies

Ringlet . . .. Adults emerge at the beginning of July and there is a peak of numbers in the middle of that month. A small number of faded individuals linger into August.

J.A. Thomas and N.R. Webb, *Butterflies of Dorset*, 1984.

10.1 INTRODUCTION

The timing of the flight-periods of butterflies is generally well known, and summary information for each species will be found in many books on the butterflies of Britain or its regions. Initially, this aspect of the monitoring data was thought to be relatively uninteresting, merely adding detail to an aspect of butterfly biology that was already well known. It took the work of others, notably R.L.H. Dennis and P.M. Brakefield, using data from the monitoring scheme, to draw attention to its interest. There had previously been little study of variation in flight-periods from year to year and from site to site, yet these factors have potential importance in determining the limits to the ranges of species and their responses to climatic change, and also have considerable intrinsic interest.

In this chapter, some characteristics of the flight-periods of British butterflies are examined, followed by a more detailed consideration of geographic and seasonal variation. Discussion of this latter aspect is based mainly on two species, one univoltine, the hedge brown, and one bivoltine, the wall. Finally, some results are described which suggest that the flight-period of the hedge brown may have become longer over the recording period.

10.2 FLIGHT-PERIODS OF BRITISH BUTTERFLIES

10.2.1 Univoltine species

The simplest butterfly flight-periods are of univoltine butterflies which overwinter either in the egg, larval or pupal stages. Examples include the spring-flying grizzled skipper and the summer-flying marbled white (Figure

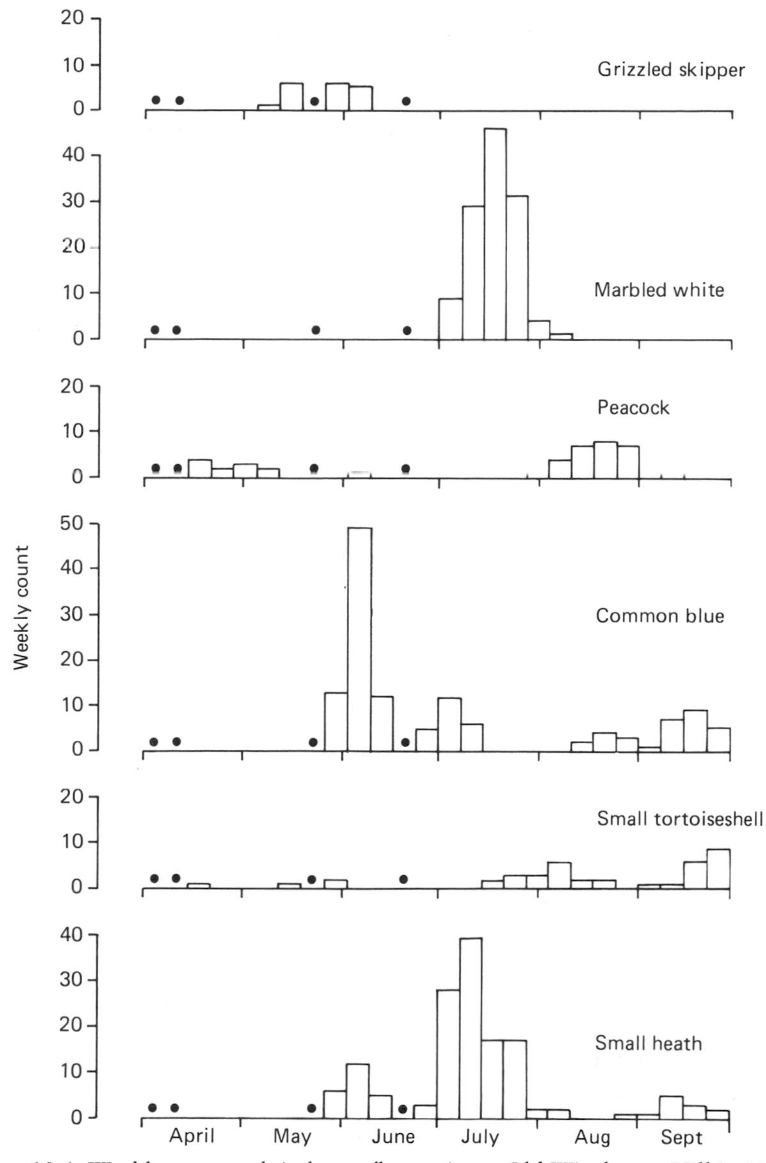

Figure 10.1 Weekly counts of six butterfly species at Old Winchester Hill in 1985, to illustrate the variety of flight periods. A dot indicates no count in that week. The grizzled skipper and marbled white have a single generation, flying in spring and summer, respectively. The peacock also has a single generation, but overwinters as an adult. The common blue and small tortoiseshell usually have two generations in the south of Britain, but the small tortoiseshell overwinters as an adult. The small heath has a complex life cycle in the south of Britain, with flight-periods of different generations often overlapping.

10.1); the former overwinters as a pupa, the latter as a larva. These single, simple, flight-periods are much the commonest amongst British butterflies and include 42 of the 55 resident species.

In the case of most of the butterflies which are univoltine and fly in the spring, there may be, in warm or very warm summers, a small second generation. The frequency varies; for example, in the case of the swallowtail at Bure Marshes, second generation individuals were recorded in 5 of 15 years (Figure 10.2), but, so far, in the case of the orange tip there have been no second generation individuals on monitored transects.

Even more rarely, some of the summer-flying butterflies may produce second generation individuals in the autumn. One very fresh, second generation white admiral was seen in September at Ham Street Woods in Kent in the hot summer of 1989. This is a rare event for this species in Britain and Frohawk (1934) recorded second generation butterflies only in 1911 and 1933.

In general, we do not have information on the sexes of the butterflies recorded on monitoring counts. In a few cases, such as the chalkhill blue at Castle Hill, the recorder has made separate counts of each sex (Figure 10.3). As a general rule, males emerge before females and this can be seen clearly in the case of the chalkhill blue. The most probable explanation for the earlier emergence of males is that it gives individual males the best chance of mating successfully (Scott, 1977; Wiklund, 1977b); females of many species mate only once, and there is an obvious premium on being the first male to find a newly-emerged female. Males of most butterfly species are smaller than females and may have had to sacrifice extra growth in order to emerge early (Singer, 1982).

Far more male than female chalkhill blues were recorded on the transect counts at Castle Hill (Figure 10.3). Such a bias may be found in counts even when the sexes are present in equal numbers (Chapter 3), because males tend to be more active and conspicuous than females. However, in the case of the chalkhill blue at Castle Hill, capture-mark-recapture studies suggested that there were indeed more males than females (R. Leverton, personal communication), although the ratio was not as extreme as indicated by the transect counts.

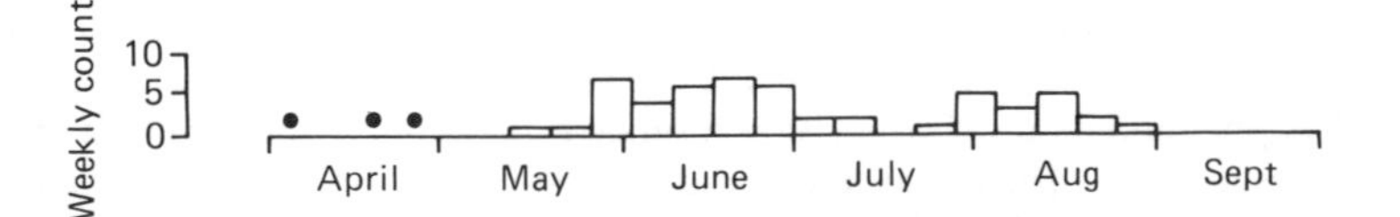

Figure 10.2 Seasonal occurrence of the swallowtail at Bure Marshes in Norfolk in 1989. In a warm summer such as in 1989, the swallowtail may have a substantial second generation in August.

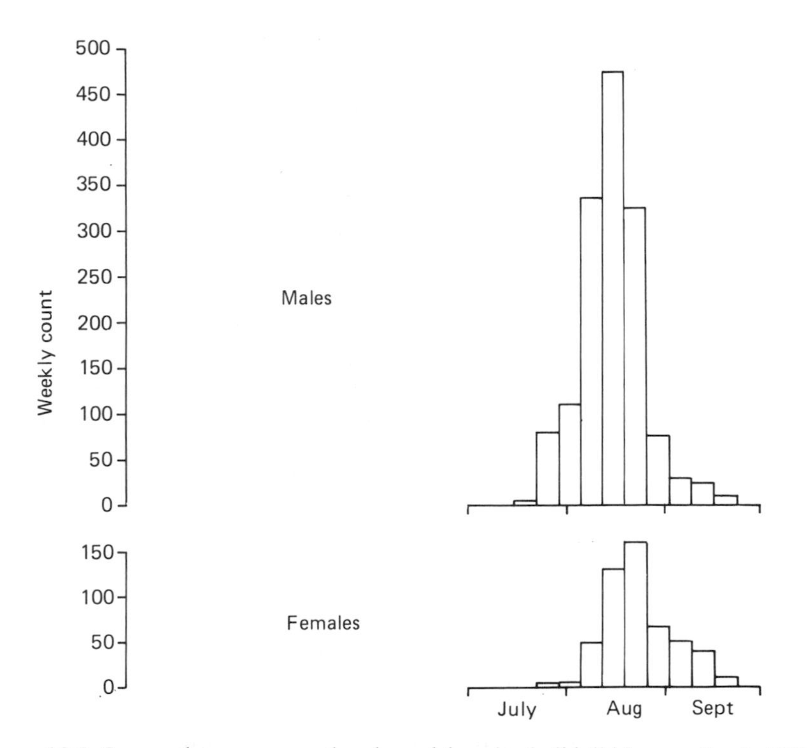

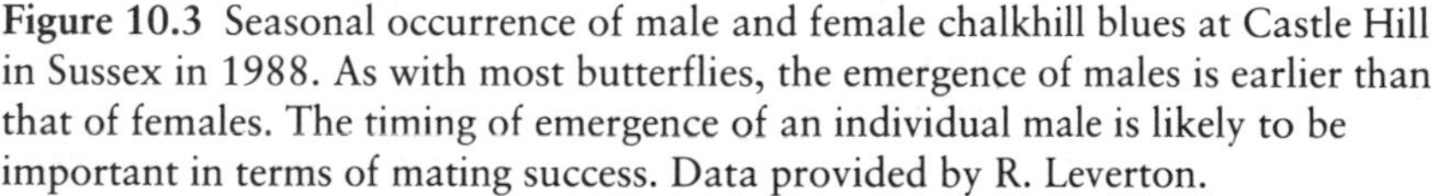

Figure 10.3 Seasonal occurrence of male and female chalkhill blues at Castle Hill in Sussex in 1988. As with most butterflies, the emergence of males is earlier than that of females. The timing of emergence of an individual male is likely to be important in terms of mating success. Data provided by R. Leverton.

Three other univoltine butterflies hibernate as adults and have flight-periods before and after hibernation. These species are the brimstone, peacock (Figure 10.1) and the (very rare) large tortoiseshell. In spring, these butterflies mate and lay their eggs, while in autumn most of their time is spent feeding at flowers prior to hibernation.

Of these three species, only the peacock is known to produce occasional second generation individuals in Britain and this species has two generations in southern Europe (Pullin, 1986). A few late September butterflies at monitored sites in the south of England in 1976 were probably of a second generation, a view that seems to be confirmed by records of late larvae in that year (Lipscombe, 1977; Holmes, 1978).

10.2.2 Bivoltine species

Eleven British butterfly species normally have two generations a year. Of these 11, nine spend the winter as larvae or pupae; examples include the common blue (Figure 10.1) and wall. Another species, the green-veined white (Lees and Archer, 1974), and also the common blue are univoltine in some northern parts of Britain. There are no monitored sites where this is clearly the case with the green-veined white, but there are several sites in Scotland and northern England where the common blue is usually univoltine (Chapter 11). At most of these sites the common blue produced a small second generation in very warm summers.

In several of these bivoltine butterflies, the second generation is partial, i.e. some of the progeny of the first generation do not become adult until the following spring, while others develop rapidly and emerge, as a second generation, in the same year. The proportion of individuals following each developmental path no doubt varies geographically and from year to year, as described below for the small tortoiseshell. In terms of analysing and interpreting population index values, this complexity presents problems. For example, successful breeding of the first generation of the green-veined white could result either in a large second generation in the current season or a large first generation in the following season, depending on temperature and perhaps other factors. We have yet to tackle this problem and our analyses ignore the partial nature of some generations.

Most of these butterflies are able to produce a third generation in warm seasons; as this generation is likely to fly at the end of September and in October, when no counts are made, the information from the scheme is fragmentary in this respect.

Two of the butterflies which are normally bivoltine, the small tortoiseshell (Figure 10.1) and comma, overwinter as adults. The overwintered individuals breed in the spring; some of the resulting larvae grow rapidly and produce adults in mid-summer and these breed again; others grow more slowly, emerge rather later in the summer and hibernate. The adults from the second generation emerge and hibernate in the autumn. Thus in both of these species the second generation is partial.

Data from the monitoring scheme on the small tortoiseshell have been examined by Dennis (1985a). He showed that a large second generation (and the occurrence of a third) was more likely in warm summers and at southern latitudes. In Scotland this species is considered to be always univoltine by Thomson (1980), but at Tentsmuir Point, a monitoring scheme site, autumn adults occasionally occur and are thought by the recorder to be locally bred (Chapter 14). Thomson's view that autumn adults are likely to be migrants from England may be generally correct, although a northward migration in autumn, presumably resulting in hibernation where winters are longer and colder, seems to be maladaptive.

The summer individuals of the comma which mate and lay eggs after emergence are brighter and have less scalloped edges to the wings than do adults which go quickly into hibernation. Presumably the shape and sombre colour of the hibernating form are adaptations for concealment amongst dead leaves or on the trunks of trees (Frohawk, 1934). It may be possible to record the two forms separately on transect counts, although, as far as we are aware, no recorder has attempted this. Nylin (1989) has shown that in the comma, the particular developmental path is dependent, not only on day length *per se*, but also on how day length changes during larval development. By manipulating day length, Nylin was able to produce the summer form of the adult in normally univoltine Swedish butterflies.

The complexity of the life cycles of the small tortoiseshell and comma has led us to use a single annual index value. There is, however, more detailed information which can be extracted from the monitoring scheme, as Dennis's (1985a) analysis shows. Separation of an index for the spring flight-period, from the later flight-periods, would add to the value of the data for these species and we may eventually do this.

10.2.3 Complex life cycles

Two of the three common migrant species in Britain, the clouded yellow and painted lady, breed continuously, without a hibernating stage. Because of the erratic arrival of immigrants, little pattern is found in the weekly counts of these species. In contrast, as described in Chapter 9, there seems to be some consistency in the steady build-up of numbers over a season, of the third of these butterflies, the red admiral.

Two resident species, the small heath (Figure 10.1) and speckled wood, can also be seen during much of the spring, summer and autumn. Both of these butterflies have complex life cycles with overlapping generations. They are able to overwinter in different stages; in the case of the small heath this may be as an early or mid-instar larva, while the speckled wood is able to hibernate as a third instar larva or as a pupa. Thus, there may be, in both of these species, staggered, but overlapping emergences in spring; a further complication is the partial development of one or two later generations. South's (1906) account of the likely succession of generations in both species is, in the main, supported by more recent studies reviewed by Shreeve and Emmet in Emmet and Heath (1989).

It is very difficult, if not impossible, to separate the long flight-periods of the speckled wood and small heath into generations and a single annual index is used in the monitoring scheme. It was in attempting to achieve such a separation of generations that led Goddard (1962) to make regular counts of the speckled wood, long before the monitoring scheme adopted transect counts for monitoring.

Thomson (1980) reported that there were at least three flight-periods of the speckled wood in most, if not all, of Scotland, while Thompson (1952) suggested that there may be only one generation in parts of Wales. There is only one Scottish site for the speckled wood in the monitoring scheme,

Taynish on the mild west coast of Argyll, although its range extends far into Scotland. At Taynish, there appear to be three flight-periods in most years and at Welsh sites such as Oxwich on the Gower, Ynis Hir on the River Dyfi and Newborough Warren in Anglesey there is a similar succession of overlapping flight-periods through the spring and summer months.

In contrast, the generations of the small heath seem to be reduced to one in many northern areas, including many of the monitoring scheme sites (Figure 10.4). Figure 10.4 suggests that, in 1990, the succession of generations was similar at the sites in southern England, but at the northern sites the timing of the single generation varied considerably. However, these examples are just a selection of sites from only one season and may not be typical; the phenological data for this species have yet to be examined fully.

Lees (1962, 1965) showed that populations of the small heath from well-separated areas of the country differed genetically in the proportion of diapausing individuals. When raised under identical conditions, individuals from southern populations were more likely to produce a second generation, rather than enter diapause. He also showed that this proportion could be changed quite rapidly by artificial selection; it seems possible therefore that, for example, a series of warm summers could cause an increase in genotypes which produced additional generations.

10.3 SEASONAL AND GEOGRAPHICAL VARIATION IN FLIGHT-PERIODS

The rate of development of insects, in the absence of diapause, is strongly influenced by temperature. The dependence on temperature of larval development, and so the timing of the flight-period from year to year, can be seen in a general way in the weekly counts of the ringlet at Castor Hanglands (Figure 10.5). The effect of warm spring and early summer, and consequently early years, of 1976, 1982 and 1989 can be seen clearly. There are many similar examples from monitored sites. Such seasonal variation in flight-periods is likely to lead to similar, or sometimes more pronounced, seasonal differences in the timing of other stages of the life cycle (as in the white admiral, Chapter 12) and may have major effects on survival.

Flight-periods' characteristics vary, not only from year to year, but from locality to locality. An extreme example is that of the meadow brown; this species has an extraordinarily long flight-period on chalk downland and on some other calcareous or dry grassland sites (Pollard, 1979b; Brakefield, 1987). In Table 10.1, sharp differences in the flight-period of the meadow brown, at a downland site in Kingley Vale in Sussex and in the nearby West Dean Wood, are shown. At Kingley Vale the data suggest that the butterflies are still flying in October, when monitoring has finished, whereas in West Dean Woods, the flight period is normally over by the end of August.

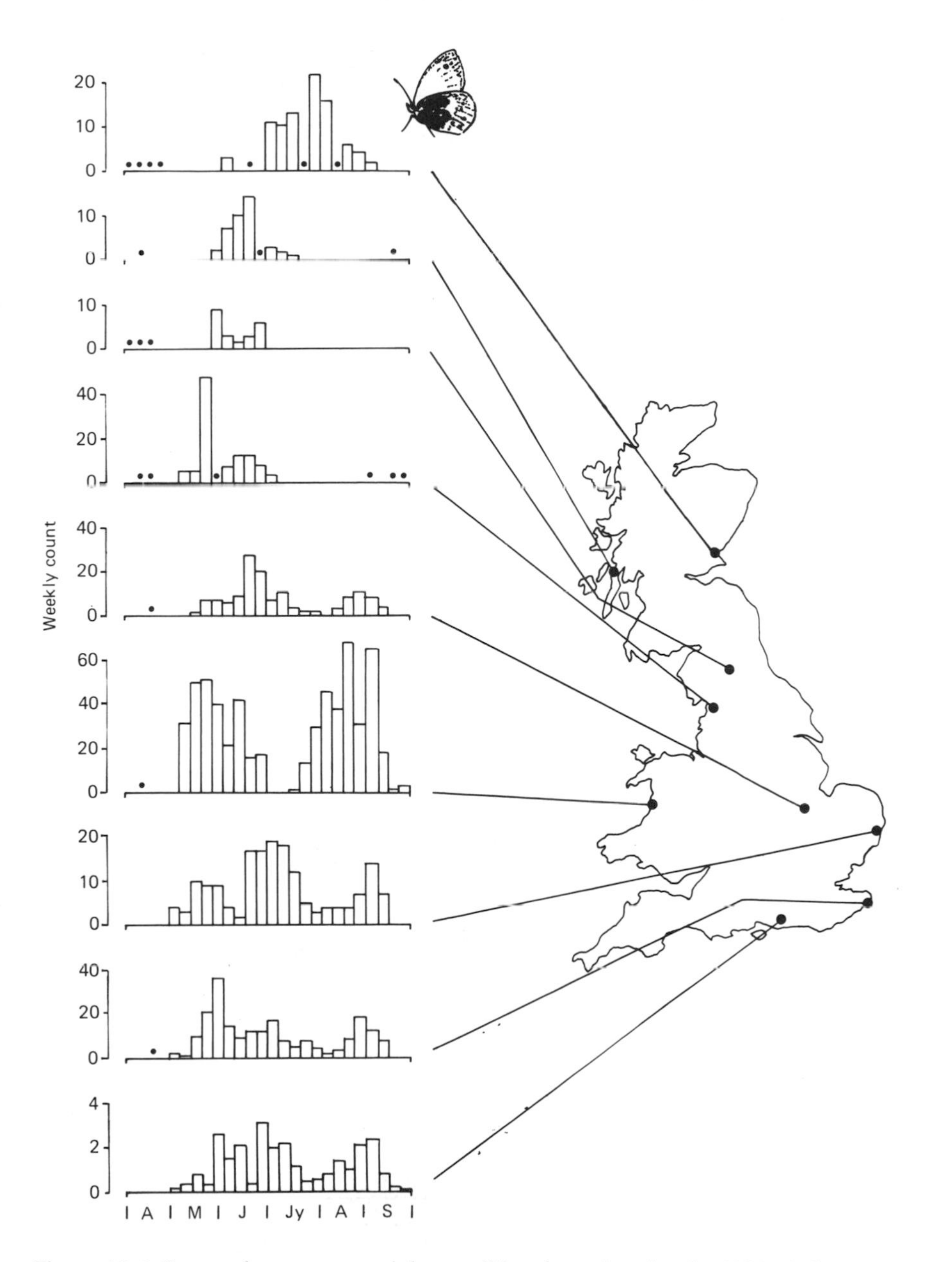

Figure 10.4 Seasonal occurrence of the small heath at nine sites in 1990. A dot indicates no count in that week; more counts were missed in the north because conditions are more frequently unsuitable for recording. In the south there is a complex series of generations. In the north, the data suggest a single generation with considerable differences in its timing from site to site.

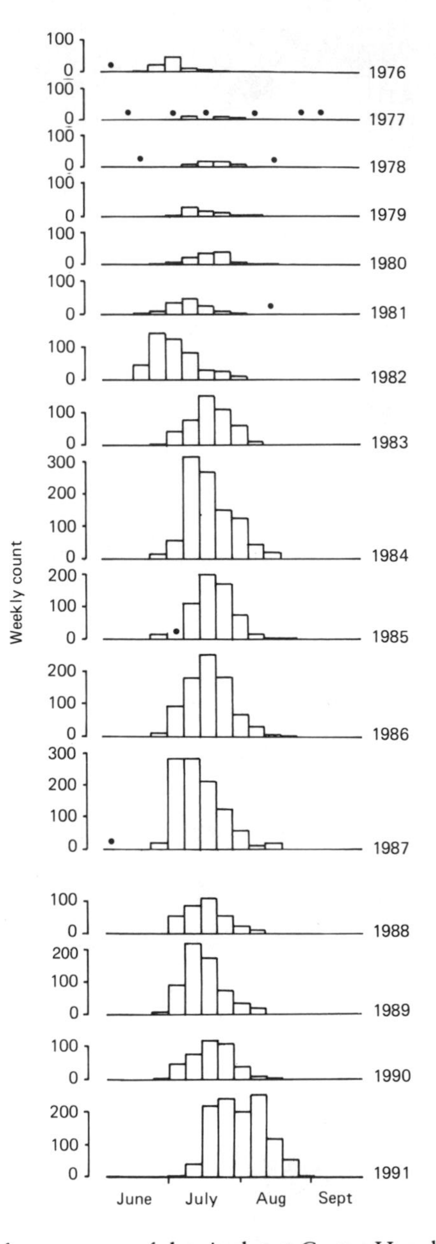

Figure 10.5 Seasonal occurrence of the ringlet at Castor Hanglands in Cambridgeshire, 1976–91. A dot indicates no count in that week. The counts show the variation in the timing of the flight-period from year to year. The ringlet increased in numbers at Castor Hanglands, and at many other sites, after a period of scarcity in the late 1970s.

Table 10.1 Differences in the flight-period of the meadow brown at two sites in West Sussex: Kingley Vale and West Dean Woods. The meadow brown usually has a greatly extended flight-period on chalk downs, and Kingley Vale is one such site. Weekly counts are given for 3 years to illustrate the extent of the differences in flight-period. A dash indicates no count in that week; spaces indicate that a count was made, but no meadow browns were seen

	Recording week number														
	12	*13*	*14*	*15*	*16*	*17*	*18*	*19*	*20*	*21*	*22*	*23*	*24*	*25*	*26*
	June				*July*				*August*				*September*		
Kingley Vale															
1983	2	7	48	91	289	411	450	473	320	217	175	42	19	5	2
1984	2	17	66	221	291	350	573	534	574	473	308	140	48	28	7
1985	–	2	8	88	148	213	311	168	221	258	146	93	89	19	16
West Dean Woods															
1983			1	20	32	46	22	22	13	1					
1984	1	3	6	22	26	39	30	27	14	10	1	1			
1985		1	3	3	22	11	9	3	3	2	2				

Shreeve (1989) has shown differences between the genitalia of early and late individuals at a downland site and suggested that two types of the meadow brown, of different geographical origins, are present.

Although the differences in the flight-period of the meadow brown at different sites is exceptional, some local variability may occur in all but the most wide-ranging migrants. The study of such variability in butterflies has only just begun, principally because no suitable data have been available before. The only published studies, using data from the monitoring scheme, are those of Dennis (1985a) on the small tortoiseshell, Brakefield (1987) on the meadow and hedge browns, and Pollard (1991b) on the hedge brown.

10.4 THE HEDGE BROWN AND WALL

In this section, some characteristics of the phenology of a typical univoltine butterfly, the hedge brown, are compared with a bivoltine species, the wall. Both of these butterflies belong to the same sub-family (*Nymphalidae; Satyrinae*) and both use a range of grass species as larval food plants. The data presented for the wall are restricted to the second generation, which flies at a similar time of year to the hedge brown, as numbers in the first generation are rather small for analysis. This account is based partly on Brakefield's (1987) study of the hedge brown and partly on our more recent studies of both the hedge brown and wall, over a longer period.

To analyse flight-period characteristics in relation to weather or to geographical trends, the timing and duration of the flight-period must be described by simple parameters. For this purpose, Brakefield (1987) used the mean flight date and the standard deviation of flight days about that date, and this procedure was followed by Pollard (1991b). Using the frequent counts of the hedge brown at Monks Wood, the standard deviation of flight days has been shown to be closely correlated with the length of the flight-period, estimated by graphical methods (Pollard, 1991b). It may seem that a direct measure of the length of the flight-period, the interval between first and last counts, would be preferable to a derived statistic such as the standard deviation. However, the interval between first and last counts is very variable from generation to generation, simply because it makes use of little of the available data; for example, it would be greatly distorted by a single individual which lived a week or more longer than the rest of its cohort.

Brakefield (1987), using monitoring scheme data for 1976–85, found that the mean flight dates of the meadow brown and hedge brown were, as expected, earlier in warm summers. Data from Monks Wood can be used to illustrate this feature for the hedge brown. June temperatures seem particularly critical in determining flight time. Following the warmest June (mean temperature 17.1°C in 1976), the mean flight date for the hedge brown at Monks Wood was 17 July, while in the coolest June (11.9°C in 1977) the mean flight date was 8 August. Brakefield also found that flight-periods tended to be shorter in warm weather, probably because emergence was more synchronized, but it is also likely that adult life is shorter at high temperatures.

The mean flight date of the wall is also earlier in warmer summers. The earliest mean date for the second generation at Monks Wood was 2 August in 1976. After the coolest June (1977), there was an index of only 2 at Monks Wood, but the mean date (22 August) was the latest recorded. For both the hedge brown and wall, the correlations between mean flight date and June temperature were significant.*

The flight-periods of the hedge brown and wall were examined in relation to latitude and longitude at sites in the monitoring scheme (Pollard, 1991b; unpublished data), although only the results for latitude are discussed in detail here. Adequate data for the hedge brown were available in every year, but only in 4 years, in which the species was most abundant, for the wall.

Given that the hedge brown flies earlier in warm seasons, one might expect it to fly earlier in the south of its range than in the cooler north in a given year. Surprisingly, there was no consistent correlation between mean flight date and latitude. In 2 years the mean date was significantly earlier in the north and in only 1 year was it significantly later. In contrast, the mean

* Hedge brown, $r = -0.73$, $P < 0.001$; wall, $r = -0.62$, $P < 0.05$; data for 1973–90.

flight date of the wall was later in the north in all 4 years; the relationship was significant in two of these years and was close to significance in a third (Table 10.2).

As the peak emergence of the hedge brown, in a given year, was more or less the same over all of its range, this implies adaptation to avoid early emergence in warmer regions. One possible mechanism for this could be the selection of cooler niches by larvae in the south; alternatively, individuals in the south may be capable of physiological delay in their growth. Bink (1985) and Wickman *et al.* (1990) have demonstrated such delays in the growth of larvae of satyrine butterflies. In contrast, the wall seems to develop as rapidly as possible and fly as early as possible; thus it flies earlier in the warmer south of Britain and is more likely to fit in an extra generation there. The wall was included amongst the species studied by Wickman *et al.* (1990). They showed that it had higher growth rates than two univoltine species, the meadow brown and grayling, both of which had reduced growth rates in the final larval instar.

This difference between the species may reflect a general difference between univoltine and bivoltine species. However, it is too early to be confident that all bivoltine butterflies behave similarly to the wall; results for the Adonis blue and common blue, discussed in Chapter 12, suggest that some bivoltine butterflies may also have delays incorporated into their development. In the case of the hedge brown, with only one generation each year, the timing of this generation seems to be particularly important, although the critical factors which make it important are not known. In the case of the wall, and perhaps other bivoltine species, success is achieved in a different way, by speed of development and a rapid succession of generations, with the penalty that unfavourable conditions are more likely to be encountered in some generations.

Although the mean flight date of the hedge brown was unrelated to latitude, the length of the flight-period was slightly, but significantly, shorter in the north (Pollard, 1991b). The same trend was evident in the data for the wall and was significant in 1 of the 4 years. Brakefield (1987)

Table 10.2 Correlation between flight time and latitude in the wall butterfly. The second generation of the wall was significantly later in the north than in the south in 2 of the 4 years in which analyses could be made, and the relationship was close to significance in a third year (1982). Significant correlations between mean flight date and latitude: indicated as follows: * $P < 0.05$; *** $P < 0.001$.

	1982	*1983*	*1984*	*1989*
Number of sites	29	32	18	21
Correlation coefficient	0.36	0.59***	0.22	0.45*

suggested that a shortening of flight-period in the north may be a response to a shorter growing season. This explanation may be true of some species, such as the hedge brown, but is unlikely to be true of the wall, as it is able to breed both much earlier and much later in the year in its first and, occasional, third generations.

A shorter flight-period carries implications for the dynamics of the populations; a butterfly with a short flight-period could experience either very bad or very good weather which lasted for the entire flight-period; thus population variability is likely to be high, with an increased chance of extinction. Some other stages of the life cycle are also likely to be more synchronized if the flight-period is short and may similarly be affected by periods of good or bad weather.

10.5 CHANGE IN RANGE AND FLIGHT-PERIOD IN THE HEDGE BROWN

The hedge brown was chosen for these studies on phenology for two main reasons. First, Brakefield (1987) suggested that the northern edge of its range may be determined by its phenology. Brakefield found that, while the flight-period of the meadow brown was markedly shorter in northern England than in the south, the flight-period of the hedge brown shortened relatively little with latitude. He suggested that this lack of flexibility may restrict its range. Second, since Brakefield's study, it has become clear that the range of the hedge brown has expanded quite considerably northward in recent decades (Figure 10.6). There is thus an opportunity to discover whether the expansion of range has, as Brakefield's study might indicate, been accompanied by changes in flight-period characteristics.

Detailed examination of distribution records of the hedge brown, held by the Biological Records Centre at Monks Wood Experimental Station suggested that the expansion of range has been in progress since about 1940 (B.C. Eversham, personal communication). There was an earlier contraction of range, dating from about the turn of the century and the post-1940 expansion has led to the regaining of much, but not all of the lost ground. A few isolated populations survived the period of retraction. Eversham estimates that the northern limit of the hedge brown, excluding the isolated populations, has moved some 50 km northwards since 1970.

Although later results, based on 35 sites over the period 1976–89 (Pollard, 1991b), support Brakefield's (1987) finding that the length of the flight-period of the hedge brown was similar over its range in Britain, the fuller data showed it to be slightly, but significantly, shorter in the north. In accord with this trend, the flight-period was therefore significantly shorter close to the edge of the range (Figure 10.7).

The next stage in this study was to discover whether the flight-period of

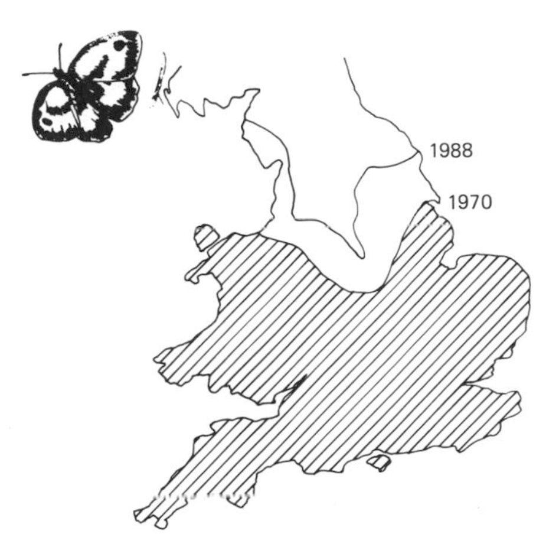

Figure 10.6 Expansion of range of the hedge brown butterfly from 1970 to 1988. The area of expansion includes some sites at which the species survived during an earlier period of contraction of range. Information from the Biological Records Centre (B.C. Eversham) and Emmet and Heath (1989). Redrawn from Pollard (1991b).

the hedge brown had changed over the recording period, while it was moving northward. As has been described, temperature plays an important role in year-to-year variation in mean flight date and in the length of the flight-period. For this reason, we tested for simple trends over time and also used multiple regression methods which took temperature into account. Again, the analysis was based on 35 sites.

There was clear evidence that the flight-period had become longer over the recording period. The trend was significant in the simple regressions and was more pronounced when June temperature was taken into account in the multiple regression analysis (Pollard, 1991b). The longest series of data, at Monks Wood, showed this most clearly (Figure 10.8). The mean flight-period for the 35 sites (which was approximately 40 days) lengthened by about 5 days over the recording period. This change seems to have occurred more or less evenly over the range. In addition, there was evidence that the mean flight date had become slightly earlier over the same period.

The apparent change in the length of the flight-period, during a period in which the range of the hedge brown has expanded, supports Brakefield's (1987) suggestion that the limits to the range may, in some way, be related to the flight-period, although perhaps not in exactly the way he thought. An increase in the flight-period could be caused by increased longevity of adults or by an extended emergence period. The latter seems more likely as

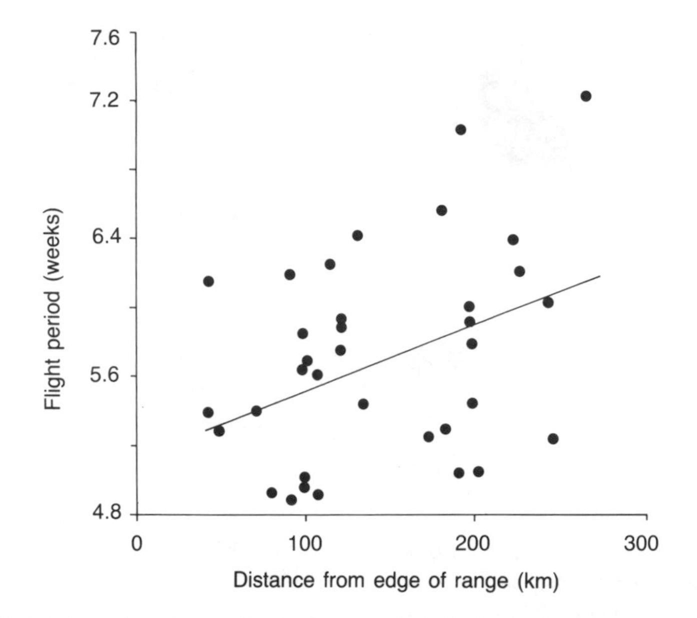

Figure 10.7 Mean duration of the flight-period of the hedge brown (approximated by four standard deviations, see text) at 35 sites, in relation to distance from the edge of the range (regression slope 0.004, $P < 0.05$). Thus there was a weak, but significant tendency for the flight-period to be shorter close to the edge of the range. Redrawn from Pollard (1991b).

increased longevity would result in a later mean flight date, whereas the actual change was to an earlier date.

As the changes in the flight-period were widespread over Britain the factor or factors responsible must also be widespread. Possible candidates include aspects of weather, other than those taken into account in the analysis, and amelioration of, or increase in, atmospheric pollutants. All of these factors could act directly on the butterflies, in one or more stages of their life cycles, or on their food plants. There have been suggestions that atmospheric pollutants may have been implicated in the declines of some butterflies (e.g. Barbour, 1986), based on correspondence between areas of high pollution and of butterfly extinctions.

Further clues on the nature of the changes in the hedge brown's flight-period may be present in monitoring scheme data. A next step might be to make similar analyses of the phenological data for other butterflies to discover whether similar trends are evident and, if so, whether there are any features common to the species affected. The changes detected in the phenology of the hedge brown could be merely an interesting curiosity or could be of some importance and have implications for the conservation of

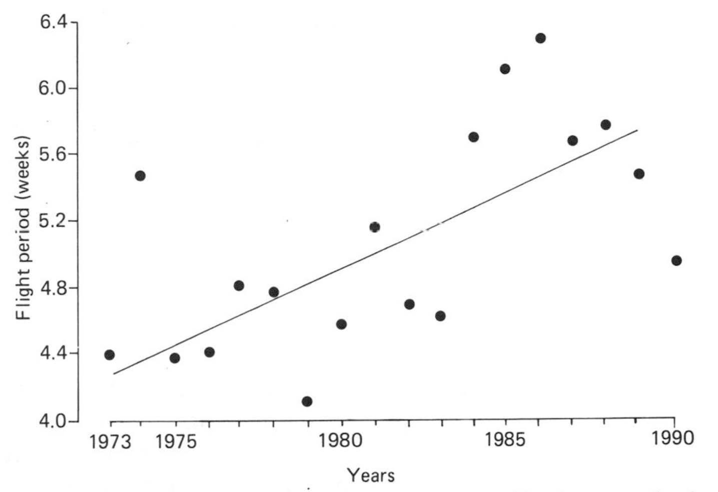

Figure 10.8 Duration of the flight-period (approximated by four standard deviations, see text) of the hedge brown at Monks Wood, 1973–90. There was a significant increase in the flight-period (regression slope 0.075, $P < 0.01$). Redrawn and updated from Pollard (1991b).

butterflies and of insects more generally.

10.6 CONCLUSIONS

The Butterfly Monitoring Scheme houses a unique body of data on the flight-periods of butterflies. As yet, there has been little analysis of these data, but the results presented here, especially on geographical and temporal trends, indicate its potential interest.

The flight-period is just one manifestation of the phenology of butterflies. The interaction between phenology and the population dynamics of species, touched on in the next two chapters, is likely to prove to be one of the major determinants of the limits of geographic range.

11

Widespread butterflies of the countryside

11.1 CHOICE OF SPECIES

In this chapter, we concentrate on four common butterflies to illustrate the range of information available from the monitoring scheme. Some more general information is also given, as background to the monitoring results, but it is not the intention to provide summaries of all that is known of the ecology of these butterflies. The selection of butterflies for more detailed treatment was not easy, as the data for all species have features of interest. The migrants were omitted, because they have been considered in a separate chapter, as also were those, such as the hedge brown and wall, which have been considered elsewhere in the book. The species chosen are the brimstone and ringlet, which have one generation each year, and the common blue and holly blue, which have two.

11.2 THE BRIMSTONE

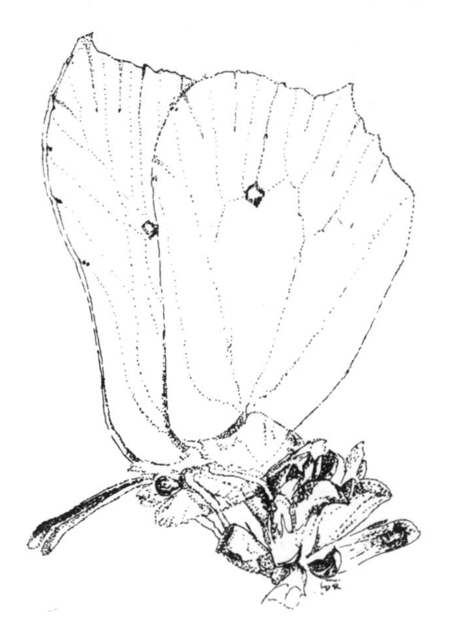

The brimstone overwinters as an adult butterfly, with individuals on the wing in the autumn and again in the spring; in the monitoring scheme, separate index values are given for each of these flight-periods. There are few reports of hibernation sites, although Frohawk (1934) states that the undersides of holly and ivy leaves are mainly used. One of us (E.P.) saw a brimstone in flight on a warm day in early March 1984; it settled under a bramble leaf about 1 m from the ground and remained in that general position (with some slight movement) until the next warm weather 42 days later. This may have been a typical overwintering situation, but perhaps more concealed sites are usual.

The eggs are laid on common buckthorn and alder buckthorn, usually in sunny situations (Bibby, 1983; McKay, 1991), often before the leaves have expanded. The females range widely to find buckthorns and can often be seen, obviously searching, flying along woodland rides and around shrubby vegetation more generally. The range occupies the southern half of Britain (Figure 11.1) and, as noted by Heath *et al.* (1984), is very similar to the combined ranges of its food plants; buckthorns and brimstones are very rare in Scotland.

The brimstone is a very long-lived butterfly and it is possible that a few individuals survive for more than 12 months. Certainly, at some sites, adults of the new generation begin to emerge in July while the last of the overwintered adults are still flying.

11.2.1 Absolute abundance

As emphasized earlier (Chapter 2), the index values from individual sites were not originally intended for comparisons of absolute abundance; the length and widths of the transect routes vary and a route may not be representative of the site as a whole. Nevertheless, major differences in index values at different sites clearly have some relation to abundance, and, used with an awareness of the limitations, provide useful information.

The largest index values of the brimstone have been at sites where buckthorns are abundant; these can be divided into wetland sites, such as Wicken Fen and Chippenham Fen in eastern England and Leighton Moss in Lancashire, and sites on calcareous soils with abundant common buckthorn, such as Aston Rowant and Kingley Vale.

Although the brimstone flies long distances, there are many sites in the scheme, within its range, where it is scarce or absent. These sites seem to be in areas of local scarcity of buckthorns and include sites near the coast of East Anglia, a wood (Northward Hill) in north Kent, chalk downland sites in Sussex and south Dorset, and Hampstead Heath in London (there have been no records at Hampstead Heath where counts have been made since 1978). Thus, although the brimstone is undoubtedly a butterfly which flies

Figure 11.1 Sites in the monitoring scheme with index values for the brimstone in five or more years.

widely through the countryside, its density is by no means uniform over its range.

11.2.2 Changes in abundance and effects of weather

Brimstone populations have been amongst the most stable, with collated index values from all sites, for the summer flight-period, varying by a factor of about three over the whole recording period. Numbers were highest in the warm years of the early 1980s. There was a particularly marked decline in abundance at monitored sites in northwest England in the mid-1980s (Table 11.1); this decline was shared by the peacock and to a lesser extent by the small tortoiseshell, although these two species recovered more quickly than did the brimstone. As discussed in Chapter 8, these declines may have been caused by the cold summer weather of 1985 (Wilson, 1991).

Table 11.1 Index values of the brimstone at two sites in northwest England, showing the sharp decline in this area in the late 1980s. A dash indicates that too few counts were made for an index to be calculated

	1977	*78*	*79*	*80*	*81*	*82*	*83*	*84*	*85*	*86*	*87*	*88*	*89*	*90*	*91*
Leighton Moss															
Spring	56	62	13	31	18	37	47	105	95	10	4	3	3	4	10
Summer	79	36	20	41	27	56	59	–	–	2	3	–	1	5	1
Gait Barrow															
Spring		41	6	12	11	5	18	56	31	4	3	1	3	5	6
Summer		24	27	23	14	80	45	67	22	3	11	0	2	8	9

As index values are calculated before and after hibernation, it is possible to separate the contributions of overwinter survival and summer breeding success to overall changes in numbers from year to year. The variability in overwinter survival is more closely related to overall year-to-year changes than is variability in breeding success.* With the usual reservation that the data are index values, rather than population estimates, the implication is that success in overwintering may be important in determining population fluctuations of the brimstone; this seems to be the first time such a finding has been reported for any butterfly.

The view that overwinter mortality is important in the population dynamics is supported by the significant association between abundance and weather before hibernation (Chapter 8). It was found that warm weather in summer was associated with increased index values in the following year. One interpretation of this association is that, in fine weather, feeding before hibernation is easier and winter survival better. Pullin (1987) showed experimentally that restriction of time for adult feeding reduced the overwintering survival of peacocks and small tortoiseshells, and it is reasonable to suppose that the same is true of the brimstone. Although hibernation success appears to be important in determining population fluctuations, there was no indication that winter

* Correlation coefficients, log transformation, 0.64 between year-to-year changes and overwinter changes and 0.43 between year-to-year changes and spring-to-summer changes; significance tests not appropriate because of lack of independence.

weather during hibernation had an effect on survival.

The spring index of the brimstone, based on all sites, is on average slightly larger than that of the previous autumn (the geometric mean summer to spring increase is × 1.2, data to 1991); it is clearly impossible for the number of individuals to increase during hibernation, thus the autumn index must be smaller, relative to the actual population size, than the spring index. The reason for this is not known, but could be related to differences in behaviour in spring and autumn, leading to differences in apparency. This bias does not affect the analysis of fluctuations as it is the variability of index values that is of interest, not the absolute levels.

A curious feature of the brimstone data is that at some sites, especially in southwest Britain, numbers recorded in summer are consistently very low relative to the spring. Results for two sites, Yarner Wood on the edge of Dartmoor and Oxwich on the Gower Peninsula in South Wales show this feature very clearly (Table 11.2). It is usually possible to think of several possible explanations for a particular unusual feature of data, but in this case we have no suggestions that seem convincing.

11.2.3 Incipient migration in the brimstone

In 1979, when processing the data for the year, it was noticed that the flight-period of the brimstone varied greatly from site to site in eastern England. In the Monks Wood area, the brimstone flew in April and May after hibernation, but then disappeared and was not seen again until the new adults appeared in July. However, at some other sites the species was recorded virtually throughout the summer.

A more detailed analysis (Pollard and Hall, 1980) showed that at woodland sites generally there was a long mid-summer gap in brimstone records, as at Monks Wood. In contrast, at wetland sites where alder

Table 11.2 Index values of the brimstone at two sites in southwest Britain, showing apparent abundance in spring and scarcity in summer, typical of the region as a whole. The reason for this is not known. A dash indicates that too few counts were made for an index to be calculated

	1976	*77*	*78*	*79*	*80*	*81*	*82*	*83*	*84*	*85*	*86*	*87*	*88*	*89*	*90*	*1991*
Yarner Wood																
Spring	38	22	43	21	17	10	26	23	66	31	18	26	29	36	41	37
Summer	11	5	5	3	2	1	4	3	5	0	4	6	6	5	7	5
Oxwich																
Spring	20	49	26	28	12	0	20	22	35	31	10	–	22	–	29	–
Summer	2	20	9	5	6	18	19	4	4	3	6	18	–	–	–	6

buckthorn was common, there was little evidence of a decline in numbers until the end of June (Figure 11.2) when, presumably, the overwintered adults died. At these wetlands in 1979, the new generation of brimstones was recorded at the end of July, but in the woods none were seen until a week or two later. There was thus an interval of about 2 months without sightings in the woods, but at the wetlands, the old and new adults came close to overlapping. Results are presented only for 1979, but the general pattern was repeated in other years.

Some eggs were laid on the common buckthorns in Monks Wood, and probably in the other woods, but the pattern of results suggests that the spring butterflies did not die in the woods, but flew out into the countryside. Monks Wood had only 12 bushes of common buckthorn in its 157 ha at this time (Bibby, 1983); of these 12 bushes, eggs were laid in any number only on the two in open, sunny, positions. It seems likely that the suitable buckthorn bushes were so sparsely distributed in Monks Wood, and perhaps in the other woods, that searching females were more likely to leave the wood rather than find one of them. Similarly, males may have left when the chance of finding females declined.

If these dispersing brimstones arrived, presumably by chance, at a wetland site with abundant food plants (mostly alder buckthorn), they were likely to remain, possibly for the rest of their lives. The wetland sites in the scheme all had areas of trees and shrubs which seemed suitable for

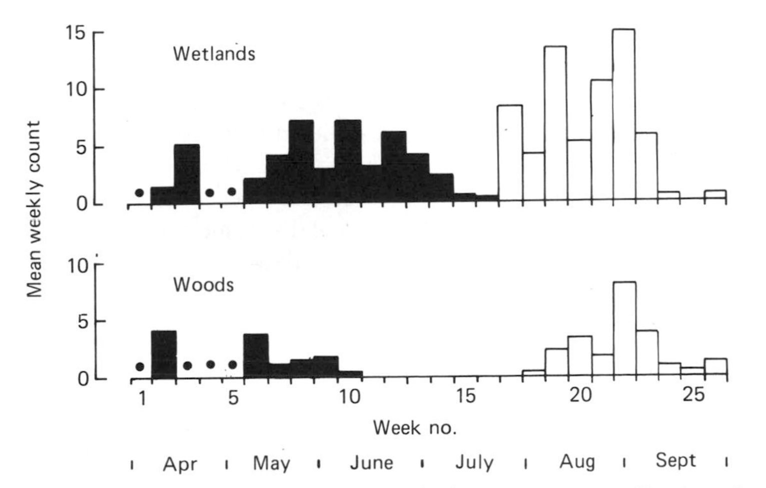

Figure 11.2 Mean weekly counts per site of the brimstone in woodlands and wetlands in eastern England in 1979. A dot indicates no counts in that week. In the wetlands the brimstone was recorded throughout the season, but in the woods there was a mid-summer gap of almost 2 months. It is suggested that the butterfly uses the woods primarily for overwintering. Redrawn from Pollard and Hall (1980).

hibernation of the brimstone, and some butterflies may have overwintered in these places. However, others may have flown away, perhaps initially to find flowers. Those that arrived in woods were likely to hibernate there. The range of these movements is not known, although there is little doubt that brimstones can fly several kilometres. It is possible that some individuals return in the spring to lay some of their eggs at the sites at which they hatched, but if so this will be by chance.

As has often happened, monitoring had drawn attention to an interesting aspect of the biology of a butterfly, in this case possible movement between hibernating and breeding areas, but could not itself take the understanding further. In the case of the brimstone, the monitoring results led to other research. Subsequently, study of egg-laying by the brimstone in Monks Wood, led Bibby (1983) to conclude that many more butterflies were seen in the wood than could possibly have been bred there. This is consistent with the use of the wood predominantly as a hibernation area. Analysis of the chemical composition of most adult brimstones caught in Monks Wood showed marked differences from those known to have bred there as larvae (Dempster *et al.*, 1986); again the conclusion was that most butterflies found in the wood did not develop there.

Even though the movements of the brimstone between hibernation and breeding sites may be the inadvertent result of flight in search of egg-sites or flowers, it is easy to see how such behaviour could lead to the evolution of one type of migration. If breeding and hibernating areas become separate for some reason, it is a fairly small further stage to the development of purposeful flights to better quality sites for either breeding or hibernation. If, as may often be the case, favoured areas for hibernation are in warmer regions to the south of the breeding sites, then directional flight may evolve. Curiously another type of, fairly local, movement of brimstones between hibernating and breeding areas has been recorded. In Lebanon, Larsen (1976) found that breeding of the brimstone (the same species as in Britain) was at high altitude, but many individuals flew to much lower levels to hibernate.

11.2.4 Current status

The brimstone seems to have been more or less stable in its range over the last 100 years (Heath *et al.*, 1984) and has been relatively stable in its abundance at sites in the monitoring scheme. The only clear exception to this generalization has been the recent decrease in abundance at the sites in northwest England, at the northern edge of its range in Britain.

The development of scrub on abandoned agricultural land could benefit this species considerably, as also could climatic warming. Thus there seems no reason, for the forseeable future, why this striking butterfly should not continue to brighten the first warm days of the year.

11.3 THE COMMON BLUE

11.3.1 Introduction

The common blue is indeed generally common throughout Britain, except at high altitude, from the south coast of England to the north of Scotland. On the continent it reaches the extreme north of Scandinavia (Higgins and Reilly, 1970) and is present virtually throughout Europe.

Hibernation is in the larval stage and the adults of the first generation usually emerge in May. Eggs are laid on a range of leguminous plants, with birdsfoot trefoil and black medick commonly used. The second generation adults fly in late summer and Dennis (1985b) recorded a switch in major food plants, dependent on seasonal availability, in the two generations. In the north of Britain, where the common blue has only one generation each year, the flight-period is mainly in July and August. The adults are thought to be quite sedentary and to occur in discrete populations, although Thomas and Lewington (1991) suggest that they wander more than their close relatives.

The common blue is one of the very few British species in which voltinism appears to be disjunct, i.e. the usual number of flight-periods switches from two to one, with a clear, although not mapped, geographical boundary (Figure 11.3). Presumably, at the boundary itself, there is some sort of intermediate voltinism, but as far as we are aware this has not been studied. The holly blue is another similar species, but the area of Britain in which the holly blue has just one flight-period seems to be rather small. The geographical separation of the two forms of the common blue seems broadly similar to that of the two, very similar, species of brown argus. The

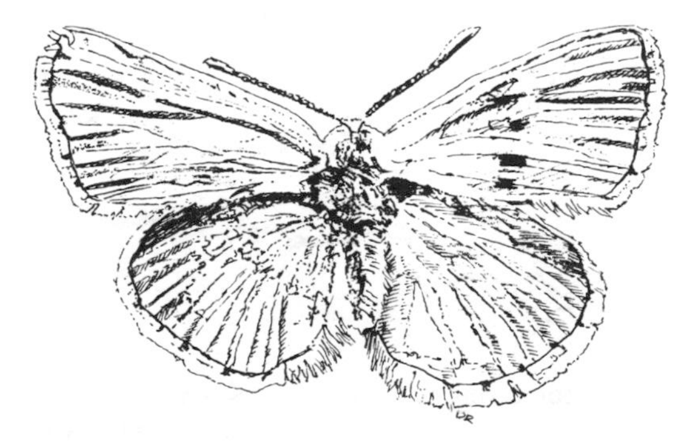

brown argus is a southern species with two generations, while the northern brown argus is univoltine.

Frohawk (1934) considered that only a small proportion of larvae from the first generation of the common blue develop rapidly and emerge as adults in the same summer, the rest overwinter and emerge in the following spring. While Frohawk was presumably correct in that the second generation is partial, the monitoring results suggest that a high proportion of

Figure 11.3 Sites in the monitoring scheme with index values for the common blue in five or more years: closed circles, southern populations with two generations each year; open circles, northern populations with one generation each year.

larvae develop quickly, as the second generation index is usually larger than the first (Figure 11.4). It is likely that the proportion breeding in the same season decreases to the north, and there is some indication of this in the selection of sites illustrated in Figure 11.7.

11.3.2 Absolute abundance

Most sites which have had high index values for the common blue are chalk downs. Nine of the 16 sites with index values of over 300 were downland sites, while the others included several with coastal sand-dunes. Two coastal sites in the north, Murlough in Northern Ireland and Newton Links in Northumberland, at both of which the common blue is univoltine, were amongst those with highest index values, with Murlough having the highest index of any site (2100 in 1989).

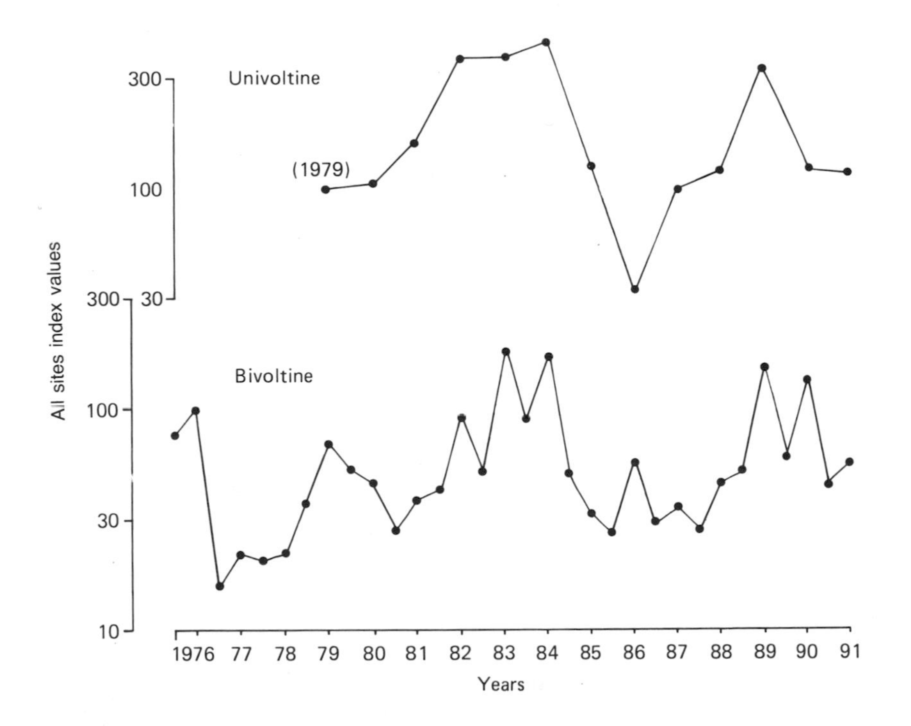

Figure 11.4 Fluctuations in numbers of northern (univoltine) and southern (bivoltine) populations of the common blue. Collated index values based on all sites; sufficient northern sites were recorded only since 1979. The indexes have arbitrary initial values.

There are no sites, with more than one year's data, at which the common blue has not been recorded. However it is rare at some farmland, woodland, wetland and upland sites. For example, at Alresford Farm, an arable farm in Hampshire, the sum of the index values for 13 years (26 generations) was 27; at Coeddyd Maentwrog, a woodland in Gwynedd, Wales, the sum was 12 in 8 years (16 generations); and at Upper Teesdale in the Pennines, 3 in 16 years (16 generations).

At these sites and at some others, numbers seen are so low and intermittent that the existence of permanent colonies must be doubted. Using our criteria for identifying extinctions and colonizations (Chapter 7), the number for this species was seven extinctions and 11 colonizations (data to 1991), suggesting that there has been more turnover of populations than for most other species at the monitored sites.

11.3.3 Changes in abundance and effects of weather

The calculation of an all-sites index for the common blue presents something of a problem because of the existence of populations with one and two generations. The sites are therefore divided into those at which the species normally has one generation and those at which it normally has two, and at the latter separate collated index values are calculated for each (Figure 11.4). Sufficient data for univoltine populations have been available only since 1979.

The two types of populations show the same general pattern of fluctuations. In the southern populations with two generations, numbers fell sharply in 1977, recovered to a minor peak in 1979 and reached higher peaks in the warm summers of the early 1980s and of 1989 and 1990. In the northern populations, the peaks coincided with those in the early and late 1980s in the south. The general similarity of the fluctuations in the north and south is rather surprising, given the very different seasons at which some stages of the life cycle occur.

Analysis of associations between changes in numbers and weather has, to date, been restricted to bivoltine populations. As with the brimstone, the generations of southern populations enable more detailed studies of associations with weather than is possible for most species. The analyses described in Chapter 8 can be extended to examine breeding success of the spring and summer generations separately, bearing in mind that this may be complicated by the partial nature of the second generation.

The increase in index values from spring to summer was strongly and significantly associated with warm and dry summer weather (Figure 11.5). Changes in numbers between the summer generation and that in the following spring were less strongly associated with weather, but nevertheless there were some clear patterns in the results (Figure 11.6). Conditions for the overwintering generation appeared to be most favourable when the previous spring was wet and the summer cool; these results suggest that dry conditions may have reduced breeding success. In contrast, high rainfall in the autumn seemed to be detrimental.

A beneficial effect of warm and dry weather in the current summer was evident in the earlier analysis of the data for the common blue, using only the second index each year (Chapter 8). However, the suggestion that dry conditions in the previous summer may be detrimental depends on the inclusion of spring and summer indexes separately. Indications of a similar detrimental effect of dry weather was found for other species in the earlier general analysis. In the case of the common blue such an effect had been suspected (e.g. Thomas and Lewington, 1991), but not shown previously in any quantitative studies. In the 1976 drought, Thomas and Lewington observed desiccation of the food plants so severe that larval survival seemed unlikely; the analysis of the monitoring data suggests that less extreme dry

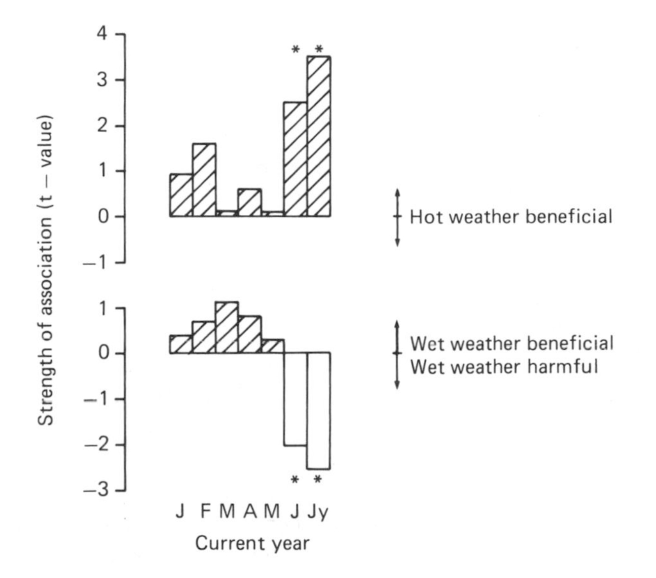

Figure 11.5 Association between spring-to-summer changes in numbers of the common blue, as shown by the all-sites index values of the southern (bivoltine) populations, and weather, 1976–90. The histogram columns show the significance of the associations (the *t*-value, considered positive or negative, for the partial regression coefficient of the weather variable, see text) with temperature and rainfall. The weather data are pooled over a 3-monthly period, e.g. the 'June' association with temperature is with the mean temperature in May, June and July. Significant *t*-values indicated by: $^*P < 0.05$. Increase in numbers from the spring to the summer generation was associated with warm and dry weather.

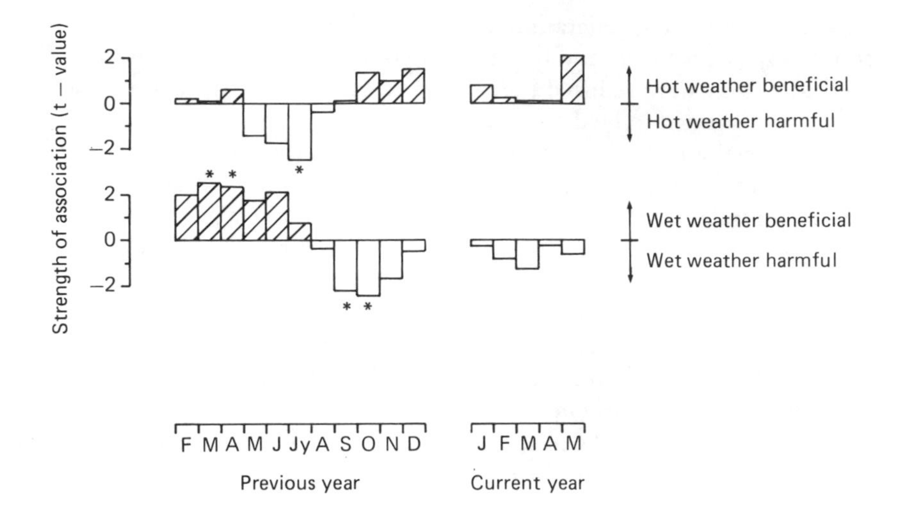

Figure 11.6 Association between changes in numbers of the common blue from the summer generation of one year to the spring generation of the next, 1976–90. Presentation as in Figure 11.5, except that weather in both years is considered. Significant *t*-values indicated by: $^*P < 0.05$. Overwinter survival was greatest when rainfall was high early in the previous year and temperature low in the previous summer. These results suggest that drought is associated with a decrease in numbers. However, there is also a beneficial association with dry weather in the previous autumn.

conditions may also reduce breeding success.

By analogy with the Adonis blue (Chapter 12), the increase in numbers of the common blue in the cool summers of the late 1970s may have been essentially a recovery from low population levels, caused, at least in part, by the 1976 drought. If, when populations are low, there are extensive areas of unexploited food plants, there may be potential for population growth even in relatively poor weather conditions.

There was no indication that winter weather was important in the overwinter survival of common blue larvae. The tendency was for warm and dry winters to be beneficial, but none of these results was significant. The lack of any strong association between winter temperatures or precipitation and changes in numbers was a feature of the more general analysis of weather and abundance, described in Chapter 8.

11.3.4 Flight-period

Some of the univoltine populations of the common blue in the north of Britain have very long flight-periods (Figure 11.7). Certainly, these single flight-periods are usually longer than either of the two in the south. In the

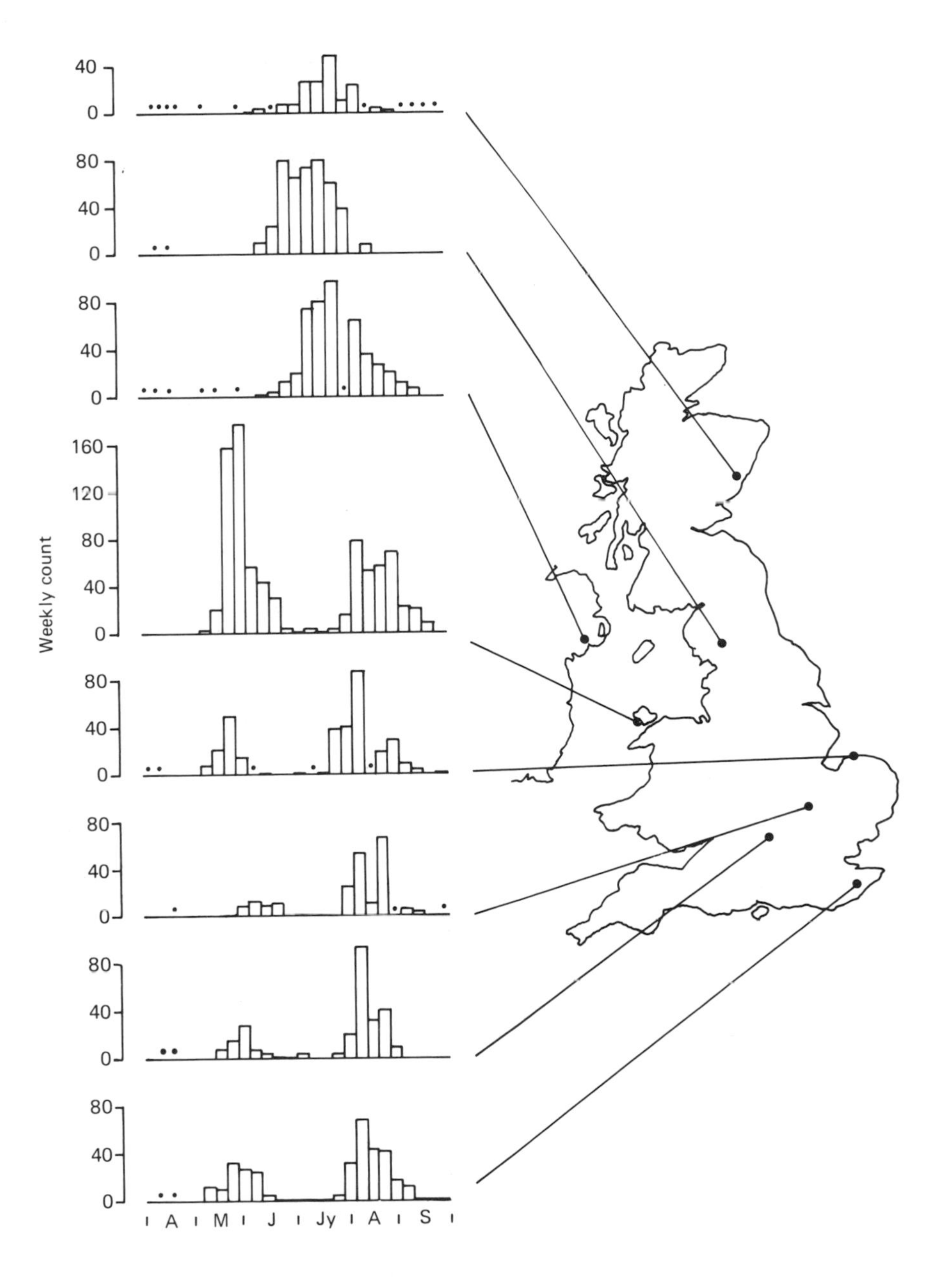

Figure 11.7 Seasonal occurrence of the common blue at eight sites in 1990. A dot indicates no count in that week. There is a clear distinction between populations with one and two generations a year, with the single generation in the north occurring in mid to late summer.

case of the hedge brown and meadow brown, the flight-period shortens towards the northern edge of the range (Brakefield, 1987; Pollard, 1991b). A similar shortening could occur much farther north in the case of the common blue, which in Britain is well within the northern edge of its range. The data illustrated (Figure 11.7) are for 1990, a very warm year, and there were a few second generation individuals at one of the northern sites. However, in this year, while the small copper and wall had clear third generations within the recording period, no third generation is detectable for the common blue at these sites, at others examined, or in other hot summers. It is likely that a small third generation of the common blue occurs occasionally in October, after monitoring has finished, but it does not seem to be a regular occurrence in this species.

11.3.5 Current status

The widespread occurrence of the common blue at monitored sites confirms its ability to breed in a variety of biotopes. Its wide range of food plants, relatively high mobility and flexible phenology, combine to make it one of the more robust survivors amongst the butterflies of Britain and Europe. Nevertheless it is believed to have lost many breeding sites in areas of intensive agriculture during this century (Heath *et al.*, 1984). If such intensification continued, this could be one of the butterflies which suffers most; equally, it is likely to benefit to some extent if more land is taken out of agriculture and left uncultivated. However, uncultivated land tends to develop a vegetation dominated by coarse grasses and scrub, and the common blue would not flourish in such conditions. There has been no indication of a decline at monitored sites, although the extent of year-to-year fluctuations is such that any long-term trends could at present be obscured.

11.4 THE HOLLY BLUE

11.4.1 Introduction

The food plants of the holly blue include a range of shrubs, and the larvae feed mainly on their flowers and developing fruits. In spring the most common food plant is probably holly and in autumn ivy is undoubtedly the main food plant, as few other native shrubby plants flower at this time. Heath *et al.* (1984) suggested that if eggs were laid amongst the flowers of male holly bushes the larvae were certain to die of starvation, because no berries were available. However, larvae have since been observed to survive to maturity on male holly, feeding on the (relatively) tender young shoots (Pollard, 1985).

The holly blue overwinters as a pupa and, as in other species, overwintering in this stage enables early emergence in the spring. Because of its association with woody plants, this butterfly often flies high in the canopy of shrubs and trees, especially in evening sunshine. It is the only blue that occurs commonly in suburban gardens.

The holly blue is scarce in the north of England and absent from Scotland, where there have been only a few isolated records over many years (Thomson, 1980). Like the common blue, there are two generations in the south of England, but only one in parts of the north; also like the common blue the second generation is said to be partial (Frohawk, 1934), with some individuals not emerging until the following spring. However, in contrast to the common blue, it has a single generation only in a restricted area. The sites in the monitoring scheme are in the northwest of England and this is also the area mentioned by Willmott, in Emmet and Heath (1989). At these monitored sites, the single flight-period is in the spring.

The holly blue does not seem to form local populations which persist for many generations. As discussed in Chapter 7, extinction and recolonization of sites, and larger areas of countryside, seem to be characteristic of its population ecology. Typically, even in years of abundance, it does not occur in high numbers at any one site in the monitoring scheme, but in low to moderate numbers at virtually all sites within its range.

11.4.2 Changes in abundance and effects of weather

This blue has long been known to fluctuate greatly in numbers (e.g. South, 1906). Even during the relatively short period of the Butterfly Monitoring Scheme, which began in 1976, there have been pronounced peaks and troughs of abundance (Chapter 6). Although the holly blue is included in this chapter on common butterflies, there have been some years in which it could be described more accurately as rare. In the summer of 1987 in particular, only three individuals were recorded at all monitored sites.

The widespread synchrony of fluctuations in the populations of many butterflies was emphasized in Chapter 6. The holly blue is an exception. Overall, the species has fluctuated widely in abundance, with peaks in 1976, 1984 and 1990 and troughs in 1981 and 1987. The all-sites index has ranged from 0.7 in the second generation of 1987 to 209 in the second generation of 1990. In all regions, there have also been three periods of abundance that are similar in general timing, but in each case the increase began in the south of England and was delayed by a year or more farther north in the midlands and East Anglia (Pollard and Yates, 1993).

During the most recent population increase, from 1989 to 1991, there was a spread from south to north in the pattern of occurrences over these years (Figure 11.8). Apart from the populations in the northwest of England with the single, spring, flight-period, the northernmost records in 1989 were in Hertfordshire and Dyfed. By 1991, there were records from virtually all sites in the scheme as far north as Lincolnshire.

At the northernmost sites (again excluding the populations of the northwest), the 1991 records were the first in many years of monitoring. For example, both Moor Farm and Gibraltar Point in Lincolnshire had been monitored since 1974, as part of a pilot study even before the start of the national scheme, without previous records of the holly blue. Evidence from other sources also suggests that the species has spread to areas which have been unoccupied for many years. For example, in a Leicester garden, the holly blue was abundant in 1991 but there had been no previous sightings in 20 years of thorough recording (D.F. Owen, personal communication).

The pattern of results suggests that, in the most recent period of abundance, there may have been a movement north by butterflies dispersing from high populations in the south. It seems that this recent spread northward could be a more extreme example of a pattern that has occurred twice before since 1976, and that part or all of the asynchrony may result from such movements.

The idea that the holly blue may spread across the countryside in periods of abundance is not a new one. At the beginning of this century, a note on the holly blue (Kaye, 1900) was published in one of the entomological journals, and included the following:

> As at many other places this year the species was noted for the first time [at Worcester Park, near Epsom in Surrey]. The sudden appearance of this insect in many places in the southern and midland (I have not heard if northern also) counties, where previously it was almost unheard of, is peculiarly interesting. One is tempted to enquire whether the species maintains an existence regularly, but in such scanty numbers as to pass unnoticed, in all these recently observed localities, or whether there has been a dispersal or migration from anywhere.

There is no doubt that the holly blue is a very mobile butterfly, as in most

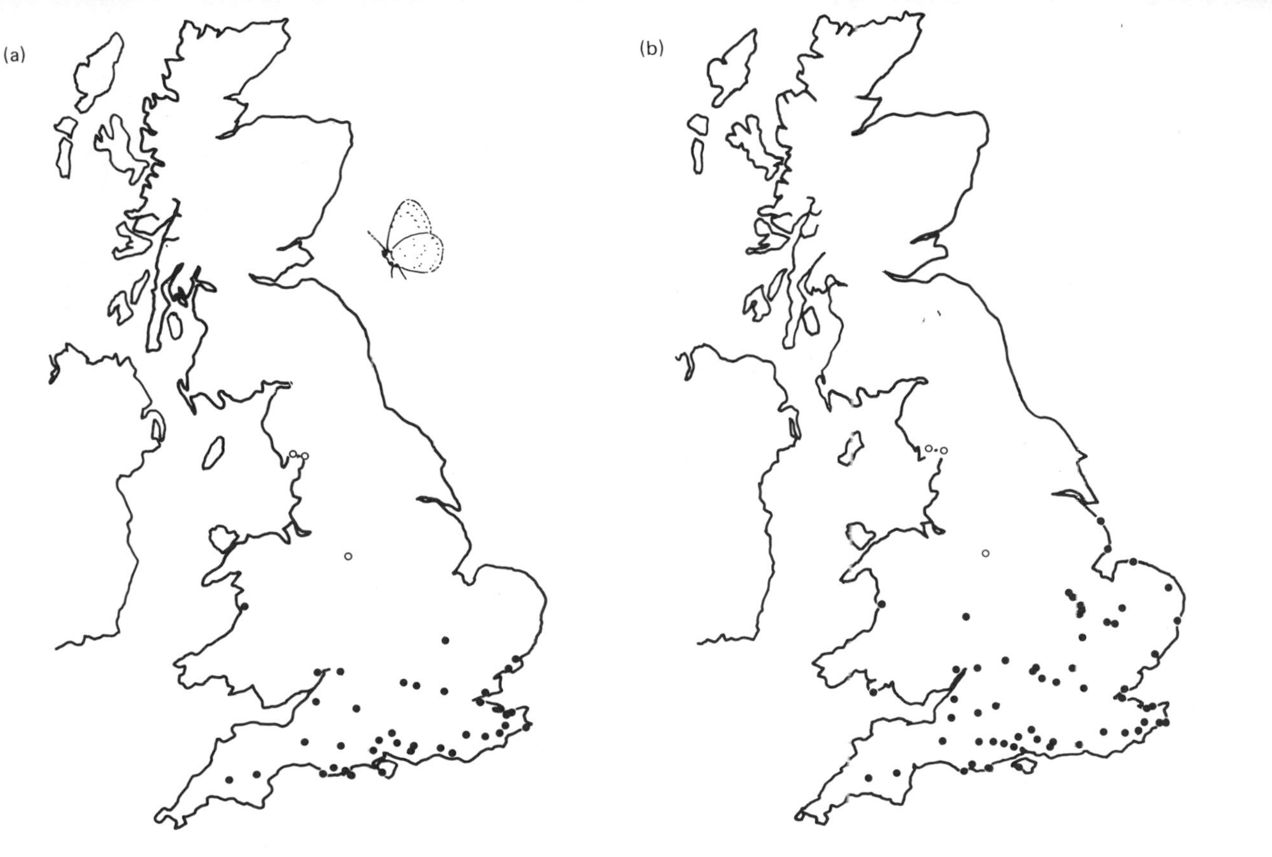

Figure 11.8 Sites at which the holly blue was recorded in (a) 1989 and (b) 1991. Between these years the holly blue increased in abundance and appeared to spread rapidly northwards. If this was the case, the rate of spread was the most rapid recorded for a resident British butterfly. Open circles indicate univoltine populations (redrawn from Pollard and Yates (1993)).

parts of the country the available flowering food plants differ in each generation and the butterflies must frequently fly some distance to locate them. However, the apparent speed and scale of movement is surprising; it is of the order required for immigration from the continent, and there is no evidence in the literature that the holly blue is a migratory butterfly (e.g. Williams, 1930).

An alternative explanation of these results (also mentioned by Kaye in the above quotation) is that the fluctuations in each region are mainly the result of local population fluctuations and that any movement of butterflies was restricted to the northern edge of the range in 1991. If this is the case, the apparent absence of the species from many sites was only the absence of sightings, in populations which had fallen to very low numbers.

The fluctuations of the holly blue since 1976 suggest the possibility that more or less regular cycles of abundance and scarcity may occur, but a much longer period would be needed to confirm this. The holly blue has a common, host-specific, parasitoid (Willmott, in Emmet and Heath, 1989) and the possibility of a host–parasitoid interaction, leading to cycling of the two populations, cannot be discounted. However, increases in numbers have generally begun in warm summers (Chapter 8), suggesting that the factors responsible for the population fluctuations are essentially similar to those of other bivoltine butterflies, although acting more dramatically in the case of the holly blue. Until evidence from detailed studies of the population ecology of the holly blue is available, discussion of the causes of both its large population fluctuations and the differences in their timing can only be speculative. We hope that these results from the Butterfly Monitoring Scheme will stimulate such studies.

11.4.3 Populations in northwest Britain

The limited monitoring data for the populations in northwest England, where there is a single annual generation, suggest that these populations may be more stable than those in the south (Table 11.3). As these northern populations are likely to have a single food plant, great mobility of individuals is not essential and thus the whole character of the population dynamics could be affected.

However, in periods of great abundance, such as in 1990–91, it is possible that individuals from the south could mix with the local populations of the northwest and influence their genetic character. There is clearly considerable scope for the study of this and various other aspects of the ecology of this fascinating butterfly.

Table 11.3 Index values of the holly blue at the three sites in northwest England where it has occurred over a long period. In most years there is a single, spring, generation. Although the data are few and the index values low, the index values seem to be more stable than in the south, where there are two generations. The periods of greater abundance correspond approximately with those in the south. A dash indicates that too few counts were made for an index to be calculated

	1977	*78*	*79*	*80*	*81*	*82*	*83*	*84*	*85*	*86*	*87*	*88*	*89*	*90*	*91*
Gait Barrows															
Spring	–	0	0	0	0	0	2	4	7	10	6	2	2	3	9
Summer	–	0	0	0	0	0	0	0	0	0	0	0	0	0	0
Wart Barrow															
Spring	–	5	1	1	–	0	1	3	2	0	1	0	0	3	0
Summer	–	–	0	0	–	0	0	0	0	0	0	0	0	0	0
Leighton Moss															
Spring	0	1	0	6	3	8	7	11	3	3	11	2	6	11	9
Summer	0	0	0	0	0	1	1	0	–	0	0	0	0	1	1
Spring totals	–	5	1	7	–	8	10	18	12	13	18	4	8	17	18

11.4.4 Current status

Although the holly blue is, at the time of writing, in a period of abundance greater than for many years, it will not be surprising if it is once again rare in 2 or 3 years. Its extreme population fluctuations have a distinct similarity with those of the black-veined white, *Aporia crataegi*, which had intermittent periods of abundance before becoming extinct in the 1920s (Heath *et al.*, 1984). We do not suggest that the holly blue is also likely to become extinct, but the possibility cannot be entirely discounted in a species which fluctuates so violently. It would be ironic if its survival depended on the relatively small number of, apparently, more stable populations in the northwest of Britain.

11.5 THE RINGLET

11.5.1 Introduction

The ringlet occurs commonly in damp sheltered lanes, amongst scrub, and along hedgerows and woodland rides, although in the north of its range, in Scotland, it also occurs in more open situations. The adults fly in July and early August, often in cool, overcast, conditions and the eggs are dropped to the ground amongst the grasses on which the larvae feed. Lear (in Emmet and Heath, 1989) gives tufted-hair grass and creeping bent as important

food plants, Thomas and Lewington (1991) give cock's-foot and wood false-brome, while Higgins and Reilly (1970) list sedges as well as grasses. Clearly, the range of food plants may be quite large, but Thomas and Lewington suggest that only one main grass species is likely to be used at any particular site. The larvae overwinter, feeding when the weather is mild (Frohawk, 1934), and complete their development in the spring.

11.5.2 Absolute abundance

Index values of the ringlet have tended to be higher at northern sites than in the south (unpublished data). This, rather surprising, correlation may occur because the sites which have the most suitable habitat conditions also tend to be northerly; in particular, several woods on heavy clay soils in Cambridgeshire (which are northerly as regards the sites in the scheme where the ringlet occurs; Figure 11.9) have large index values, while the species is generally scarce, or even absent, on the chalk downs and heaths farther south. There have been no records at all in 14 years at the downland reserve of Castle Hill in Sussex and no more than 10 over an even longer period at Ballard Down near Swanage. However, where woodland or scrub is well developed on chalk soils, the species may be moderately abundant. Thus at Old Winchester Hill the ringlet has had index values of between 10 and 100 in most years, but has been largely confined to the edge of a small wood (see also the account of the butterflies on the downs at Wye in Chapter 13). At another site on chalk, Gomm Valley in Buckinghamshire, where scrub is well developed, the ringlet is common throughout most of the area. Heathland sites usually have few records, probably because, like the open downs, they are hot and dry. There have been no records of ringlets at Studland Heath in Dorset in any of the 16 years of recording and very few at Walberswick in Suffolk over the same period.

Figure 11.9 Sites in the monitoring scheme with index values for the ringlet in five or more years.

Although the relationship between size of index values and latitude may have more to do with the nature of the sites than latitude *per se*, it is interesting that the ringlet has particularly high index values at Morton Lochs and Tentsmuir Point (Chapter 13), by far the most northerly sites for the species in the monitoring scheme. At these sites, as elsewhere in Scotland, the ringlet occurs in damp grassland in the open, and even in relatively dry conditions.

In favouring more open situations in the north, the ringlet contrasts with the speckled wood, another butterfly which is predominantly associated with shade and humidity. The speckled wood tends to fly more widely in the open in the south and west of Britain (e.g. Thomas and Lewington, 1991), while in Scotland Thomson (1980) associates it with shady woodland. The difference is intriguing, but we can offer no explanation.

Warren (1985) and Greatorex-Davis *et al.* (1992) used transect counts to show the preferred shade conditions of the ringlet and other species in woodland in southern Britain (Chapter 5). Not surprisingly, the ringlet was one of the butterflies most tolerant of shade, with high counts in as much as 50–60% shade. As the ringlet is believed to be a very sedentary species, these results should relate to its breeding areas, not just to areas of adult flight.

11.5.3 Changes in abundance and effects of weather

The collated index values based on all monitored sites (Table 11.4) show that numbers fell sharply in 1977, rose steadily from 1977 to 1982 and subsequently remained relatively stable. There has been a significant increase, assessed over the whole recording period, but this occurred mainly during those first few years. There is some evidence that the largest increases have been in eastern England. The synchrony of fluctuations of the ringlet over its range in Britain has been particularly close (e.g. Pollard, 1991a), suggesting that weather plays a dominant role in its population ecology.

The ringlet is one of those species we thought to be most susceptible to drought (Pollard, 1988); now we are less sure. As discussed in Chapter 8, the statistical relationship with drought for this species depended largely on the sharp fall in index values following the extremely hot and dry summer of 1976, followed by an increase during several succeeding wet summers. It is possible that only a very exceptional drought, such as that of 1976, has a major impact on numbers; certainly there seems to have been little or no adverse effect of further, fairly severe droughts in 1989 and 1990. An adverse effect of drought on the ringlet is very much in accord with its known preference for damp sites, but the relationship is obviously more complex than it first appeared.

11.5.4 Colonization and extinction

There has been recent expansion in the range of the ringlet in Scotland (R. Leverton and M.R. Young, personal communications) and in the north of

Table 11.4 Collated index values of the ringlet, based on all sites. The index starts at an arbitrary value of 100 in 1976

1976	*77*	*78*	*79*	*80*	*81*	*82*	*83*	*84*	*85*	*86*	*87*	*88*	*89*	*90*	*91*
100	33	59	102	146	168	325	317	404	301	358	372	264	300	225	369

England. In some cases, the movement has been from the north, filling in gaps in the distribution (P.K. Kinnear and S. Ball, personal communications). However, there has been no 'recognized' (see Chapter 7) colonization of new sites in the monitoring scheme; unfortunately, we have no sites in the areas of range expansion in the north. However, it is reasonably certain that the island of Skomer in South Wales has been colonized in recent years. The first record on the monitored transect was in 1980, the fourth year of recording. Skomer lies only about 1 km from the Welsh mainland, but as the ringlet is generally considered to be a very sedentary species (e.g. Thomas and Lewington, 1991), colonization of this type is presumably an unusual event.

Although no extinctions from monitored sites have been recorded, it seems likely that the ringlet disappeared from a field in Monks Wood for a period of several years, after the 1976 drought (Pollard, 1984). The failure to recolonize the field, when numbers built up considerably elsewhere in the wood, was surprising. It seems unlikely that the adults were so sedentary that they failed to move back into the field, and perhaps more likely that a particular food plant suffered from the drought and took several years to recover its abundance.

11.5.5 Flight-period

As in the case of the hedge brown (Chapter 10), the flight-period of the ringlet (as indicated by the mean flight date) seems to be as early in the north of its range as in the south. For example, it flies just as early at Morton Lochs in Scotland as at Alresford Farm in Hampshire (Table 11.5).

The length of the flight-period was consistently longer at Morton Lochs than at Alresford (Table 11.5). If there is more generally a longer flight-period in the north of Britain, this would be in contrast to the hedge brown and meadow brown (Chapter 10). However, a full analysis of geographical variation of the flight-period of the ringlet has yet to be undertaken.

At Monks Wood the flight-period of the ringlet was shorter than that of the other common satyrine butterflies, the meadow brown and hedge brown. Using an estimate of the length of the flight-period (Pollard, 1991b) over the 19 years of recording, the mean flight-periods were as follows: meadow brown, 55 days; hedge brown, 36 days; ringlet, 32 days. The mean flight-periods of the hedge brown and ringlet were significantly different.* As is the case at most sites, the flight-period of the meadow brown was so much longer than those of the other species that statistical tests to demonstrate this were considered unnecessary.

* Paired t-test, $t = 2.27$, $P = 0.04$.

Table 11.5 Flight-period of the ringlet at Alresford Farm in Hampshire and Morton Lochs in Fife, Scotland. Weekly counts for 3 years are shown, as illustrations of the data. In the 11 years in which there are data from both sites, the mean flight date for both sites is 18 July. The flight-period is significantly longer at Morton Lochs (paired *t*-test, $t = 5.00$, $P < 0.001$). A dash indicates no count in that week; spaces indicate that a count was made, but no meadow browns were seen

	Recording week number									
	12	*13*	*14*	*15*	*16*	*17*	*18*	*19*	*20*	*21*
	June				*July*				*August*	
Alresford Farm										
1989		24	75	132	21	2				
1990		3	9	24	64	13	1			
1991				7	33	40	15	4	3	
Morton Lochs										
1989		58	65	241	84	40	16	2	–	
1990	6	37	122	76	–	21	10	5	–	
1991			72	163	121	84	54	31	1	

11.5.6 Current status

The ringlet seems to be flourishing; it has increased in abundance at monitored sites and expanded its range in the period of recording. At some woodland sites in the east of England it has, in recent years, had the highest index values of any species, whereas in the late 1970s those of the hedge brown and meadow brown were very much higher.

There is no obvious reason for the recent success of the ringlet; it is possible that it is favoured by the absence of management and the consequent increasing shade of many woods, but the variety of sites at which increases have been recorded suggest that some more general factor, or factors, must be responsible.

11.6 CONCLUSIONS

Information from monitoring has been presented for just four of the 30 or so butterflies that are well represented at monitored sites. Although a range of topics has been considered and some interesting and unexpected results obtained, it is clear that, even for these four species, the data have not been fully explored. Some points of interest are the subject of current study, and others remain to be tackled in the future. More generally, as the theoretical basis of ecology develops, it will often be possible to re-examine the existing data in different ways to test these new ideas.

12

Rare and localized butterflies

12.1 CHOICE OF SPECIES

In England, the rarer butterflies are mostly characterized by their restriction to particular island biotopes within the matrix of the agricultural countryside. Individual populations may be large, but, nevertheless, when a species becomes restricted in this way to a limited number of more or less isolated localities it must be regarded as endangered. The actual number of localities is not known with complete certainty for any of our species (except in the case of the re-introduced large blue and large copper), but in several cases the number of sites is fewer than a hundred.

For the common butterflies, full use can be made of the synoptic nature of the monitoring scheme, and results at different sites and in different regions can be compared and contrasted. For the rarer species, this cannot be done to the same extent, although in some cases data are available from a small group of sites. In this chapter, some results of monitoring four of the rarer British butterflies are discussed and placed in context by reference to other recent studies.

Far more is known of the habitat requirements of the rarer butterflies than of common species. This is largely because, on account of their rarity, the species have been more fully studied. An additional reason may be that their requirements are more restricted, hence their rarity, and so can be defined more precisely. The requirements of the different species are described in some detail here, although the studies on which they are based are not all at monitoring scheme sites. The findings are important to an understanding of the population ecology of butterflies and so to interpretation of monitoring results more generally.

The butterflies discussed in this chapter include one very rare butterfly, the heath fritillary, a scarce downland species, the Adonis blue, and two butterflies that are more widespread, the wood white and white admiral. The three woodland species are considered first, grouping them together because trends in woodland management in the recent past are thought to have affected all of the species, although in different ways.

12.2 THE WOOD WHITE

12.2.1 Introduction

There are few sites in the monitoring scheme where the wood white is present. However, studies of this species by M.S. Warren (Warren, 1984, 1985; Warren *et al.*, 1986) have included transect counts, based on the monitoring scheme methods, as an important part of the work.

The wood white flies mainly in May and June, with a frequent, but partial and small second generation in southern localities. The larval food plants are primarily meadow vetchling, bitter vetch and birdsfoot trefoil; the particular species used depends largely on local availability. The larvae leave the food plants to pupate, usually in early September, on a stem of a robust plant such as bramble, and there spend the winter. Although predominantly a woodland species in Britain, there are some populations amongst scrub and tall grassland.

12.2.2 A population study

In a study of the wood white from 1977 to 1984 in one of the woods comprising Yardley Chase in Northamptonshire, Warren *et al.* (1986) assessed annual fluctuations in abundance and also longer-term trends under a range of shade conditions. The study was in a conifer plantation on an old woodland site; the rides retained a rich herbaceous flora, including abundant meadow vetchling, the main food plant of the wood white in this area. The rides varied considerably in the extent of shade, as the forestry trees in different woodland compartments varied in age from 7 to 45 years at the start of the study. Shading of rides was measured using hemispherical

photography as described by Evans and Coombe (1959) and, specifically as used in studies of butterflies, by Warren (1985). Capture-mark-recapture studies gave estimates of population size in some years and index values from transect counts were used to assess year-to-year changes in abundance.

The most important factors determining annual fluctuations in numbers (Warren, 1984; Warren *et al.*, 1986) were egg-laying success and survival of very young larvae. These factors were both influenced by weather. When the spring and early summer was warm and dry, the number of eggs laid was high, survival of young larvae was good and so numbers of butterflies in the next year increased; in cool and wet weather there were few eggs, poor survival of larvae and numbers fell. Similar correlations with weather in the previous summer were found with two other species which fly in the spring, the dingy and grizzled skippers (Chapter 8). As discussed earlier, it is not known whether the same factors were operating with these species as with the wood white, but this may well be the case.

On all rides the annual fluctuations were similar, but longer-term trends varied according to the amount of shading of the rides by adjoining trees. The optimum shade conditions for the wood white were found to be in the range 20–50% shade. Thus in very open rides, numbers of wood whites recorded on transect counts began at a low level and tended to increase over a period of several years, while, in rides which were at or beyond the optimum conditions at the start of the study, numbers declined.

Abundant food plants were present in the more open rides, but relatively few butterflies flew in them and virtually no eggs were laid. When eggs were transferred to plants in one of these open rides, the larvae survived perfectly well. Thus the selection of less open situations for egg-laying seems to be related to the growth form of the plants, rather than to their suitability as food. In open situations, meadow vetchling normally occurs as a relatively low-growing component of the herbaceous vegetation; in shadier conditions, especially where there is abundant ride-edge scrub, it assumes a more scrambling form and grows through, and projects above, the surrounding vegetation. The females seemed to lay preferentially on these projecting food plants, hence the abundance of eggs in moderately shady rides. The reason may be simply that these stems are more likely to be encountered by searching females than are those growing low amongst a herbaceous sward. Warren *et al.* (1986) confirmed that it was possible to increase the number of eggs laid, by removing vegetation from around meadow vetchling plants and so exposing them to searching females. Thus, as has been found with many butterflies, a large proportion of potential food plants is effectively unavailable to the wood white.

12.2.3 Extinctions and colonization at monitored sites

At Waterperry Wood the wood white has come close to extinction after a period of abundance (Table 12.1), possibly because of increased shading beyond the optimal range. There has been some clearance of trees in this wood, but it may be a few years before newly open rides have developed to a stage suitable for the wood white. Several woods on the borders of Oxfordshire and Buckinghamshire are monitored; all of these woods were planted, some years ago, with a high proportion of conifers. At Shabbington and Oakley Woods the wood white has declined in a manner similar to that at Waterperry, but at Whitecross Green Wood the population has continued to flourish (Table 12.1). The reason for the relative success of the wood white at Whitecross Green Wood is not known for certain; however, the recorder (Mrs R. Woodell) believes that infrequent cutting of ride edges, and resultant growth of low scrub, has benefited them. This view is consistent with our understanding of the requirements of the wood white.

At Monks Wood, the wood white was introduced in 1984, survived for some 4 years, but now appears to be extinct (Chapter 7). This species has recently expanded its range in Northern Ireland (Warren, in Emmet and Heath, 1989), where it occupies more open sites than in England, and has colonized the monitoring scheme site at Murlough. It was first seen there in 1981, the third year of recording, and numbers were quite large (index 44); it seems reasonable to assume that a few individuals were present, but not seen, in 1980 and bred successfully in that year.

12.2.4 The future

In recent years, especially in the 1960s and 1970s, there has been some expansion in the range of the wood white, following a decline earlier in the century. Warren (1984, and in Emmet and Heath, 1989) attributes the recovery to the conversion of many lowland woods to conifer plantations in

Table 12.1 Index values of the wood white in four woods near the Oxfordshire–Buckinghamshire border. All of the woods contain mixtures of conifers and hardwoods and are on sites of ancient woodland. –, Present, but insufficient data for the calculation of an index

	1976	*77*	*78*	*79*	*80*	*81*	*82*	*83*	*84*	*85*	*86*	*87*	*88*	*89*	*90*	*91*
Waterperry	87	90	28	19	10	0	5	2	2	1	2	0	0	0	1	0
Shabbington									14	11	1	0	1	2	1	–
Oakley									24	7	1	0	0	3	3	0
Whitecross Green											39	43	58	69	123	46

the 1950s and 1960s, but regards it as only a temporary phase, before decline is resumed. Many of the plantations were on the sites of ancient woodland with abundant food plants. As in the study at Yardley Chase described above, the rides of conifer plantations are, for a time, very suitable for the butterfly. The management of these rides is often particularly appropriate for the wood white as the scrub at the ride edge is cut at intervals of approximately 5 years (Warren, 1984), allowing scrub to develop, but not grow too tall. As these plantations have become more shaded, there is evidence, including that from the monitoring scheme mentioned above, that this phase of expansion may have ended. The immediate prospects for the wood white do not therefore seem to be good, although it is fairly widely distributed, especially in Ireland, and not in danger of extinction in the immediate future.

12.3 THE WHITE ADMIRAL

12.3.1 Introduction

This beautiful woodland butterfly is usually present at about 15 sites in the monitoring scheme each year. These sites range from Yarner Wood, Devon, in the west to Ham Street Woods, Kent, in the east and Monks Wood near to the northern edge of its range. This account is to a large extent based on a study of the population dynamics of this butterfly in Monks Wood (Pollard, 1979a), one of the northernmost sites in the scheme for the species. The study was an attempt to understand the striking changes of range which have occurred during the last 100–150 years. The white admiral is a very strong flier and capture-mark-recapture studies to obtain estimates of

population size would have required exceptional speed of foot and athleticism. Transect counts were therefore used as an alternative, to give information on relative abundance.

The adults usually fly in late June and July; the eggs are laid on honeysuckle and this is the sole food plant. The sites selected for egg-laying are in areas of dappled shade at the edges of woodland rides or in woodland with partial tree-cover. Vigorously growing plants in full sunshine are usually avoided. The young larva feeds in a very characteristic manner in the autumn; it leaves the mid-rib of the leaf uneaten and uses it as a resting site between periods of feeding. Hibernation is as a third instar larva, in a hibernaculum formed from a rolled leaf. The leaf used is usually one on which the larva has been feeding, and is secured in place on the honeysuckle stem by a silk binding. In the spring, the larva changes colour from brown to green and then feeds more generally amongst the honeysuckle foliage.

In the early 1900s the white admiral was restricted to southern England, especially Hampshire and adjoining counties. It spread during the 1930s and 1940s as far north as Lincolnshire and has more or less maintained this distribution to the present. It has proved possible to reconstruct details of the timing of the expansion from information in contemporary entomological journals (Pollard, 1979a); because the white admiral is a large and striking butterfly, its change in range attracted the attention of naturalists and, as new woods were colonized, notes and comments were frequently published. The main period of expansion appears to have been between 1930 and 1942.

There is evidence of further recent expansion, both in the west and in the east of England. In the monitoring scheme, the most striking example of colonization was of the Bure Marshes area of the Norfolk Broads. Here, in the thirteenth year of recording, it was first recorded on the transect route in 1988 and has been present in the three subsequent years with maximum index values of 9 in 1989 and 1991.

12.3.2 Changes in abundance and effects of weather

In the 6-year population study at Monks Wood, Pollard (1979a) found a significant correlation between the index of abundance and June temperature. In June the larvae usually reach their final instar and the pupae are formed. Exclusion of birds, by caging the larvae, suggested that they were the main predators of these stages. In warm weather, development is rapid and the larvae and pupae are available for predation by birds for a shorter period, hence survival is good and numbers of adults increase. June temperatures were also particularly high during the period of expansion in the 1930s. From these two lines of evidence it seemed reasonable to suppose that warm June weather led to unusual abundance and enabled the spread of the white admiral to new areas of the country.

Unfortunately for this satisfyingly tidy explanation, in the longer period of data from the monitoring scheme the association between June temperature and abundance has been weak and non-significant, although the positive relationship for June is stronger than that for any other monthly temperature (data to 1991). Not surprisingly, perhaps, given the relatively short period of the earlier study, the effects of weather may not be as simple as was first suspected. There are no strong relationships between index values of the white admiral from the monitoring scheme and weather variables that we have examined; the strongest is an association between dry weather in early spring and increased abundance, but there seems no obvious biological reason why this should be so and it may be a chance correlation. The spread of the white admiral was probably related to warm summer weather, as suggested by Pollard (1979a), but the evidence must now be considered less strong than indicated by the earlier study.

12.3.3 Habitat requirements

Pollard (1979a) suggested, on the basis of the habitat requirements of the white admiral as observed during the population study, that the spread of the species may have been influenced by changes in woodland management as well as by June temperature. Although the white admiral is seen most often in sunny woodland rides, the egg-laying females seek out more shaded areas to lay their eggs. We do not know the reason for the choice of honeysuckle in shady situations, but by analogy with the wood white and with other species, we can speculate that it is related to the particular way in which the females search for oviposition sites. Unlike the wood white, the white admiral appears to use sight for the initial location of honeysuckle leaves. Experience of searching for white admiral eggs and larvae suggests that leaves which stand out clearly against a plain background, such as bare ground or a tree trunk, are often selected; it is likely that the butterfly first identifies the characteristic shape of a honeysuckle leaf from a distance and confirms the identification after alighting. This procedure is probably typical of butterflies generally (Wiklund, 1984), although in the case of the wood white the females seem to alight on vegetation before they determine whether or not it is a suitable food plant (Wiklund, 1977a; Warren, 1984). The difference between the species seems to lie in the nature of the vegetation in which the food plants are found; in the case of the white admiral, honeysuckle frequently grows in isolation from other plants; in the case of the wood white, the food plants are usually supported by, and entwined around, a wide variety of plant species and detection from a distance would almost always be difficult.

It is possible that white admiral larvae can survive well on honeysuckle in situations in which no eggs are laid, as was the case with the wood white. However, the necessary transfer experiments, as described for the wood white, have not been made.

The conditions of shade required by the white admiral are scarce or absent in regularly coppiced woodland. Frequently the hanging stems of plants which have grown high up a tree trunk are used. Such conditions might be found where standard trees are grown amongst the coppice, but honeysuckle is regarded as a weed of woodland managed for timber and often has been removed. As many woodlands became more shaded after the decline in coppice management in the early years of this century, conditions for the white admiral must certainly have improved. The contractions of range of several butterfly species (including the heath fritillary and wood white discussed in this chapter) have been attributed, at least in part, to the decline in coppice management of woodland. In the case of the white admiral, it seems that the spread of the species was the result of a combination of favourable weather and the increasing availability of woodlands more suited to it.

Many of the butterflies associated with coppice woodland benefited, for a period, from the clearance and subsequent planting of some of these woods with conifers. With the white admiral, no such immediate benefit can be expected as the required conditions take some years to develop. Indeed it seems very unlikely that conifer plantations ever provide ideal conditions for the white admiral, as the partial, dappled shade that it requires for egg-laying will not develop. However, small populations can survive in conifer plantations, for at least a few years, with the obvious proviso that forestry management does not include the removal of honeysuckle. In Bevills Wood, a conifer plantation adjoining Monks Wood, the white admiral was first seen in 1982 (Table 12.2). This was the seventh year of recording and about 20 years after the conifers were planted on this ancient woodland site; it was recorded in several subsequent years but now seems to be extinct again. One would expect the white admiral to disappear from the wood as the conifers grow taller and shade out the honeysuckle within the woodland compartments.

12.3.4 Flight-period

The flight-period of the white admiral, like that of other butterflies with one annual generation, appears to vary relatively little over its range in Britain in any one year but varies considerably from year to year. In the case of this species, because a detailed life-table study has been made, we can examine the consequences of the year-to-year variation in weather on the timing of other stages (Figure 12.1).

The study included 1976 and 1977, years which had summers with

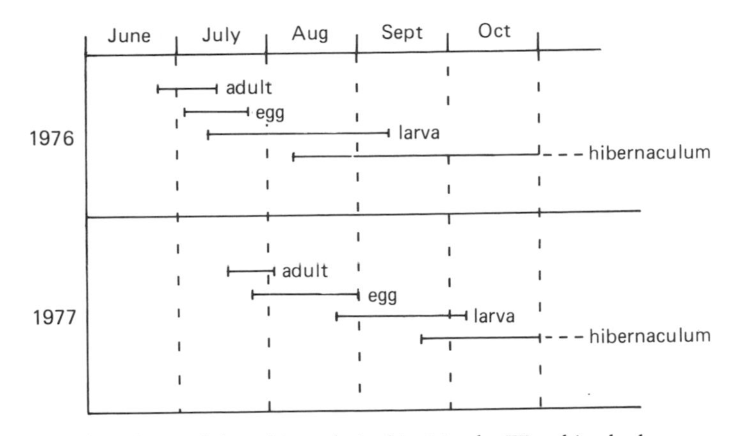

Figure 12.1 Phenology of the white admiral in Monks Wood in the hot summer of 1976 and the cool summer of 1977. First and last records of each stage of the life cycle, until hibernation, based on frequent visits to examine marked individuals. September and October of 1977 were relatively warm, otherwise some of the larvae may not have reached a stage at which they could form a hibernaculum (see text) before the onset of winter.

sharply contrasting temperatures. In 1976 at Monks Wood the mean temperature over June, July and August was 17.7°C, while in 1977 it was 14.2°C. In 1976, the adults flew early and the eggs and young larvae developed rapidly; the larvae hibernate in the third instar, and in 1976 the first hibernaculae were formed in early August and the last in early September. The summer of 1977 was cool throughout; adults flew 2–3 weeks later than in 1976 and hibernation of the larvae was not complete until well into October. It was fortunate for these larvae that the autumn of 1977 was fairly warm. Hibernation in 1977 was late, but no later than in two other cool summers in the study period, 1972 and 1974; if the autumn of 1977 had been cold, it is likely that some larvae would not have reached the stage at which they could hibernate before the onset of winter, and they would presumably have died. Thus the effect of weather on timing of the life

Table 12.2 Index values of the white admiral in Bevills Wood, Cambridgeshire. The wood is a conifer plantation on the site of ancient woodland. Our interpretation of the results is that the wood provided suitable conditions for the white admiral for a short period when the conifers supported trailing honeysuckle amongst their foliage; subsequently honeysuckle was largely shaded out within the forestry compartments. The conifers were planted in the early 1960s. A dash indicates insufficient counts for the calculation of an index.

1974	*75*	*76*	*77*	*78*	*79*	*80*	*81*	*82*	*83*	*84*	*85*	*86*	*87*	*88*	*89*	*90*	*91*
0	0	0	0	0	0	0	0	1	9	3	7	3	–	3	0	0	0

cycle may have important implications for survival of a species, especially if, as in the case of the white admiral, it is at the northern edge of its range. The effects could be sufficiently important to provide an additional reason for the spread of this butterfly in a period of warm summers.

12.3.5 The future

Available evidence suggests that the white admiral is in no imminent danger of extinction in Britain. There are still many neglected (in a forestry sense) woods which provide suitable conditions for it, and it is possible that it will continue to extend its range. As with any species at the northern edge of its range, a series of very cool summers could lead to a decline; however, climatic changes are predicted to be in the opposite direction and warmer summers are expected to speed the spread of this handsome butterfly.

12.4 THE HEATH FRITILLARY

12.4.1 Introduction

The monitoring scheme now has four sites at which this very rare butterfly has been recorded, although two have been included only recently. This account draws heavily on the work of M.S. Warren, who has studied the ecology and conservation of the heath fritillary extensively (Warren *et al.*, 1984; Warren, 1987a, b, c, 1991). Warren used various methods, such as capture-mark-recapture and study of larval survival, and also made extensive use of transect counts. An account of this species is included here partly because of its intrinsic interest and partly because the nature of its population fluctuations differ from most of the species considered, and so are of particular relevance to an understanding of the population ecology of butterflies.

The heath fritillary is now restricted to two areas in Britain, in southwest

England and in Kent. It was once quite widespread over southern England but has been declining during the twentieth century. Because of its vulnerability, we follow Warren (1987a) in not naming the specific current locations.

In Kent, where the heath fritillary is a woodland species, the larvae feed on the annual herb common cow-wheat. In the southwest, in heathland, and in the few woods where it occurs, populations also use common cow-wheat, but in grassland ribwort plantain and germander speedwell are the main food plants. The adult flight-period is usually in June and July (but is earlier in the west: see below), with just one generation each year. The eggs are laid in large masses, sometimes on the host plants but more usually on another nearby plant, and the larvae feed gregariously in a web; by the third instar most have become solitary. Overwintering is in this instar, rolled up in a dead leaf in the leaf litter. The adults are believed to be very sedentary and nearly all movements of marked individuals recorded by Warren were under 150 m, although there were very occasional flights of up to 1500 m.

12.4.2 Changes in abundance

Typically, the heath fritillary occurs in small, discrete populations, in which the number of individuals often fluctuates sharply over a short period of years. In contrast to most of the species that have been considered, the fluctuations of adjacent Kentish populations seem to be quite asynchronous; while one population may be increasing, sometimes dramatically, another nearby may be decreasing equally quickly.

Warren (1987c) has shown that the fluctuations of these populations are related to the time of coppicing (or other type of clearance) of woodland. Populations usually reach a peak 2–3 years after coppicing and then decline rapidly. The reason is that, following coppicing there are sunny conditions and often a flush in the abundance of food plants, but then shade develops rapidly again. In Warren's studies,populations became extinct only 5 years after coppicing and the species survived only if newly coppiced areas were available close by. Where coppice woodland had been cleared and planted with conifers, the butterflies survived a few years longer but permanent extinction was then virtually certain because of the absence of nearby clearances. It is almost certain that the general decline of the heath fritillary over the last century has been caused by the reduced frequency of woodland clearance by coppicing.

The population at a monitored site in Kent has shown a characteristic rapid rise and fall (Table 12.3). At this site, a succession of woodland clearings is maintained to provide suitable habitats for the heath fritillary. However, there is an element of unpredictability in any such management and, in this wood, cow-wheat has not developed abundantly in the recent clearings; probably for this reason, the heath fritillary has come close to extinction. At other sites in Kent, similar management for the species has been extremely successful.

At the few woodland sites in southwest England the population fluctuations are of a similar character to those in Kent and the butterfly responds rapidly to woodland management. However, at grassland sites in the west the populations appear more stable. These grassland populations showed some synchrony of fluctuations and Warren (1987c) found significant correlations with spring weather. This may be another example of warm weather speeding larval development and also leading to increased survival, as has been suggested in this chapter for the white admiral. In spring, the heath fritillary larvae spend much time basking in the sun, a clear indication of their need for warmth. Thus, the grassland populations of the heath fritillary more closely resemble other butterflies that we have considered, in that annual fluctuations are influenced by the weather.

The difference between the woodland and grassland populations seems to lie in the speed with which the biotopes change. In particular, coppice woodlands move rapidly from completely open conditions to more or less complete shade and the effect of these changes on the butterflies clearly overwhelms the potential effects of weather. In contrast, the grasslands change relatively slowly, even when they are unmanaged, and so effects of weather may be evident. Nevertheless, if management of grassland is neglected over a period of years, Warren has shown that there is a slow reduction in the carrying capacity of the site and a decline of the butterfly populations. The initial results from experimental management suggest that autumn mowing, combined with disturbance of the ground to encourage the food plants, will prove effective in maintaining numbers.

12.4.3 Flight-period

Warren (1987a) found that the heath fritillary emerges some 2–3 weeks earlier in the southwest of England than it does in Kent. The monitoring scheme has had a site in each area for many years and we found that there were sufficient data for comparison for 6 years, from 1983 to 1987. Over

Table 12.3 Index values of the heath fritillary in a Kentish woodland. Cleared areas have been created for the butterfly in recent years, but unfortunately the food plant, crested cow-wheat, has not flourished. The speed of change is typical of this butterfly

1982	*1983*	*1984*	*1985*	*1986*	*1987*	*1988*	*1989*	*1990*	*1991*
19	34	38	140	141	14	5	4	1	0

this period, the mean flight date in the southwest was 25 June and in Kent it was 10 July, a difference of just over 2 weeks. The difference in dates was statistically significant.*

A difference in mean flight date of some 2 weeks is large. In an earlier chapter (Chapter 10), studies on the flight-period of the hedge brown were described. The hedge brown, like the heath fritillary, emerges later in the east of the country than in the west. However, the difference is quite small; hedge brown sites as distant from each other as the western and eastern sites of the heath fritillary had differences in mean flight date of only 3 days. The reason for the large difference in flight dates of heath fritillary populations is not known. Warren (1987a) suggests that it may simply reflect differences in spring temperatures in the two areas; it is equally possible that time of emergence may be related to the availability of food plants suitable for oviposition and for the development of larvae, and that there may be genetic differences between populations in the two areas. The transfer of overwintering larvae between regions and comparisons of emergence times might resolve this point, but it would be difficult to justify a rather academic experiment in the case of such a rare butterfly.

12.4.4 The future

The heath fritillary was almost certainly the rarest British butterfly (excepting the re-introduced large blue and large copper) when Warren began his studies in 1980. His research showed the need for very active management, especially of its woodland sites. Such management has been implemented and has been very successful in a number of cases. Warren (1991) reports that the number of known colonies increased between 1980 and 1989 from 31 to 43. This increase was largely because of the discovery of new colonies in heathland on Exmoor, but many populations elsewhere have been maintained or enhanced by management. At two sites, populations have been re-established by transfer of butterflies to areas prepared by appropriate management. In contrast, most unmanaged sites have lost their populations.

The prospects for the survival of the heath fritillary as a British butterfly cannot be regarded as good, because the number of populations is small and most or all of their sites require continuing special management for the species. However, the prospects are certainly much better than they were a decade ago and the improved status represents a success for both research and conservation.

* Paired t-test, $t = 3.64$, $P < 0.05$.

12.5 THE ADONIS BLUE

12.5.1 Introduction

In sharp contrast to the rare woodland butterflies considered above, the Adonis blue is a butterfly of open countryside and is restricted in Britain to the warmest slopes of closely grazed chalk downland. It has a southerly distribution and most of the British populations are on south-facing slopes. The vivid blue Adonis blue males are seen and counted much more frequently than the chocolate brown females, as is also the case with the common blue considered in the last chapter.

The Adonis blue is moderately well represented at monitored sites, considering its rarity. There are data from six sites which span six or more consecutive generations. This account is largely based on the researches of J.A. Thomas (1983b); much of his study of this butterfly was conducted at the monitoring scheme site at Ballard Down, Swanage, and the transect counts formed an integral part of this work.

The caterpillar of the Adonis blue, like that of the chalkhill blue, feeds on horseshoe vetch. Hibernation is as an early to mid-instar caterpillar (Frohawk, 1934); the first generation adults fly in May and June, while those of the second generation usually fly in late August and September and a few individuals may survive into October.

Thomas (1983b) noted that the larvae and pupae were almost always attended by ants, which milk them for nutritious secretions. Thomas and Lewington (1991) describe the remarkable 'songs' by which both larvae and pupae communicate with the ants. The benefit to the larvae is protection from predators and parasitoids, and the ants may in this way be essential to the survival of the butterfly.

12.5.2 Changes in abundance and effects of weather

Morris and Thomas (in Emmet and Heath, 1989) estimated that populations of the Adonis blue at Swanage have fluctuated between approximately 1000 and 100 000 adults between 1976 and 1984 (range of index values

from 13 in the spring generation of 1977 to 1287 in the autumn generation of 1983). Thomas and Merrett (1980) and Thomas (1983b) suggested that the dramatic decline in numbers in late 1976 was caused by the severe drought of that summer. The food plant, horseshoe vetch, has a large woody rootstock and the plants survived the drought but their leaves were desiccated. Thus the chalkhill blue, which shares the same food plant but overwinters in the egg stage, was not susceptible to the drought. The index value of the chalkhill blue declined at Swanage in 1977, but only to a relatively small extent (index values for 1976, 505; for 1977, 328). Thomas and Merrett (1980) drew attention to this contrast between the Adonis and chalkhill blues at Swanage. They also made the more general point that butterfly species which fed as larvae during the period of the 1976 drought suffered more severe declines than the chalkhill blue and also the dark green fritillary, which enters hibernation as a larva immediately after hatching and before feeding.

Three, largely separate, populations were present on the Swanage site and occupied areas of 5 ha, 5 ha and 11 ha respectively, with distances of about 100 m between them. Capture-mark-recapture experiments showed frequent recaptures within each discrete population, but there was no recorded movement between populations. All of these areas were managed differently from 1976 to 1982, with management varying from no grazing to heavy grazing by cattle (and some ponies). The trends in index values in the three populations were very different, although the general patterns of annual fluctuations were similar. The different trends could be related to management; in general, absence of grazing led to a decline in numbers, while numbers increased sharply where grazing was heavy and the sward kept short. The importance of a short downland turf, suggested by the trends in index values of the different populations, was confirmed by the distribution of eggs and larvae. Nearly all eggs were laid in areas of turf of 4 cm height or less, where temperatures at the soil surface were high. Although closely grazed areas are necessary for the Adonis blue in Britain, extreme over-grazing in a period of drought brought it close to extinction at Castle Hill in 1980 (Pollard and Leverton, 1991). However, the species did survive and increased rapidly in numbers when grazing pressure was relaxed a little. In spite of this example, there is no doubt that undergrazing is a much greater threat to the species than over-grazing. Thomas (1983b) presented and reviewed a considerable amount of evidence showing that the Adonis blue has been lost from many sites where grazing was light or absent, and where the sward had grown well above the 4 cm required for optimal conditions.

Following the 1976 drought, the Adonis blue increased steadily at Swanage over a series of summers with rather poor weather. This seems surprising in a species which appears to select sites where temperatures are high and which is at the northern edge of its range in Britain. Thomas

(1983b) suggested, and we agree with his interpretation, that the increase occurred because numbers had fallen to a low level after the 1976 drought and there was subsequently a large amount of highly suitable, but unoccupied, habitat. Thus growth rates were high until the carrying capacity was again approached, probably in 1982, and in this period of basically unsuitable weather the butterfly was able to increase in abundance.

At the Swanage site, the fluctuations in index values over the period for which there are full data are closely correlated with the collated indexes of the common blue, based on all sites (Figure 12.2). As yet, there has been no analysis of associations between index values of the Adonis blue and weather conditions, but this correlation between its abundance and that of the common blue strongly suggest that the two species are affected similarly by weather. In Chapter 11, it was shown that the strongest effect of weather on the common blue was its increase in abundance in warm summers, while a detrimental effect of drought in the previous summer was also evident. Both factors also seem to be important to the Adonis blue (in spite of its increases in the poor summers following the 1976 drought) with an indication that its numbers fluctuate more widely than those of the common blue. The main difference between the two species was that the Adonis blue declined after the first generation in 1976, but the common blue declined

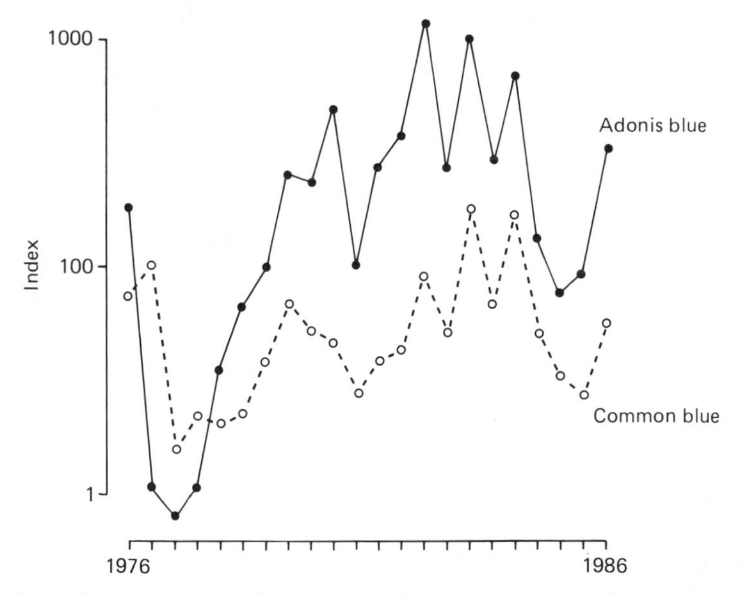

Figure 12.2 Fluctuations in index values of the Adonis blue at Ballard Down, Swanage, from 1976 to 1986, compared with the all-sites index of the common blue. The fluctuations are closely correlated ($r = 0.68$, $P < 0.001$), suggesting that the two species are similarly affected by weather.

after its second generation, further suggesting a more severe effect of drought on the Adonis blue.

The introduction and subsequent extinction of the Adonis blue at Old Winchester Hill was described in Chapter 7. At another downland nature reserve, Martin Down in Hampshire, the species has been more successful (Table 12.4). In the early years of recording the Adonis blue was present at Martin Down only in very low numbers and indeed may have been extinct for a period. Subsequently, the more general periods of abundance in the early and late 1980s have been reflected in the local results at Martin Down. However, there has been an underlying trend to increased abundance and, although still not very numerous, it now seems to be firmly established. Areas of Martin Down have been managed by intermittent intensive grazing, using sheep in small paddocks. It appears that conditions for the Adonis blue have been improved substantially. Results from additional transects, which have been recorded since 1982, seem to confirm this increase in abundance (information from P. Toynton).

Table 12.4 Gradual increase in index values of the two generations of the Adonis blue at Martin Down in Hampshire. The site has in part been managed by intermittent grazing with sheep in small paddocks

	1979	*80*	*81*	*82*	*83*	*84*	*85*	*86*	*87*	*88*	*89*	*90*	*91*
1st generation	1	0	0	0	3	6	24	5	4	0	0	13	20
2nd generation	0	0	0	4	16	21	11	8	6	10	20	41	31

12.5.3 Flight-period

A comparison of the flight-periods of the common and Adonis blues on the downs at Swanage proved both interesting and puzzling. In the first generation the mean flight date of the Adonis blue was slightly earlier (Adonis blue, 12 June; common blue, 15 June).* However, in the second generation the timing of the mean flight dates was reversed, with a difference of nearly 2 weeks (Adonis blue, 5 September; common blue, 24 August).†

The difference in flight times in spring is surprising, as the common blue hibernates in the third instar and the Adonis blue in any of the first three instars (Frohawk, 1934). Given equal development rates the common blue should emerge first; the inference is that the Adonis blue develops more quickly. However, the flight times of the second generation indicate that the Adonis blue caterpillars develop more slowly than those of the common blue during the summer.

* Difference significant at $P = 0.02$, paired t-test.
† Difference significant at $P = 0.001$.

These results suggest that there is a 'delay' incorporated into the development of the larvae of one or both of these species. The development of the common blue in the spring may be delayed, perhaps to synchronize with the period of maximum growth of the food plant, and/or the development of the Adonis blue may be delayed in the summer. The Adonis blue is close to the northern edge of its range in Britain and is restricted to the warmest slopes of chalk downs; these considerations suggest that its larval growth rate is as rapid as possible and that delayed growth by the common blue in the spring is more likely. However, without studies which address the question directly, this conclusion must remain speculative.

12.5.4 The future

Many Adonis blue populations have become extinct during this century and, until recently, few new populations appeared to have been established (Thomas, 1983b). Thomas estimated that in 1981 only some 70–80 populations survived. Since then, Thomas and Lewington (1991) suggest that the number of populations may have doubled, partly as a result of improved management of downland nature reserves, while other factors may have been an increase in the number of livestock (Thomas, 1990) and rabbits, resulting in more extensive areas of closely grazed sward and bare ground.

In spite of its recent recovery, the Adonis blue remains at risk. The number of populations is few and these fluctuate violently, with an ever-present risk of extinction. In general, population fluctuations of the bivoltine butterflies seem to be less stable than those of the univoltine species (Chapter 6). However, most of the other bivoltine butterflies are mobile species which are able to recolonize sites after extinction. The Adonis blue combines exceptional population instability with low mobility (Thomas, 1983b) and so is especially vulnerable.

Although the Adonis blue seems to have been severely affected by the 1976 drought at its existing sites, climatic warming could prove to be beneficial. The species has a clear requirement for warmth, even though the precise reason is not understood. There is little doubt that a warmer climate would extend the range of biotopes that the Adonis blue could occupy, enabling it to breed in longer swards and on downs with cooler aspects, as it does in more southern parts of its range. This butterfly does not extend as far north in Britain as its food plants and, almost certainly, a warmer climate would enable it to spread to some of these unexploited areas.

12.6 CONCLUSIONS

Each of these four accounts of rare British butterflies has been based largely on detailed research. Although our butterflies are still amongst the most threatened in Europe, their conservation has undoubtedly benefited from these studies. The contribution of monitoring to the study and conservation of the rare British butterflies has been relatively modest, but monitoring *per se* has important functions. The recommendations based on research usually entail continued management of sites; only if the populations are monitored over a long period can one be sure that the management is achieving its ends. In addition, long-term monitoring can provide tests for some of the conclusions reached in the more intensive short-term studies and generate ideas for new research. Finally, monitoring ensures a continuing interest in the fate of populations after the risk of imminent extinction has, perhaps temporarily, been averted.

– 13

Site studies

13.1 INTRODUCTION

The wide variety of sites in the monitoring scheme, in terms of topography, climate and the range of their biotopes means that the selection of a few sites for more detailed treatment cannot do them full justice. Nevertheless, we hope that the descriptions of the sites in this chapter will help to transform 'counts' and 'index values' into the real world of butterflies and their habitats.

First, the general methods used to assess changes in abundance of butterflies at individual sites, especially in relation to wider trends, are described. Then six of the sites, their butterflies and their conservation problems are described.

13.2 ASSESSMENT OF THE EFFECTS OF HABITAT CHANGE

To assess the effects of habitat change, the following assumptions have been made:

1. Annual fluctuations in index values at regional and national (all-sites) levels are caused mainly by variations in weather conditions. The longer-term trends shown by these collated index values provide some sort of average of the trends at individual sites.
2. At individual sites, departures from wider trends in collated index values are frequently the result of local changes in the conditions required by a particular species. These departures may be sudden, due to drastic management or other radical change, or gradual, often due to successional changes in vegetation.

In addition to comparisons of site data with wider trends, within-site comparisons are also informative. Often the distribution of butterflies around a transect route remains remarkably constant from year to year, but changes in local distributions sometimes occur and are likely to reflect changes in the composition of vegetation or the degree of shading, or both.

As comparisons between and within sites form the basis of assessments of, and advice about, the success or failure of management for butterflies on nature reserves, objective tests of local departures from wider trends are desirable. We use one such test for comparing long-term trends at individual sites with wider trends. In this test, the departure of the site index from wider trends is assessed, a significant departure indicating an increase or decrease relative to wider trends*.

There is usually only the most general information on changes in vegetation of monitoring scheme sites, and this limits interpretation of local departures in butterfly numbers from wider trends. However, it is sometimes obvious that sudden local changes are related to radical change in the management of a site, such as the felling of trees in woodland or the introduction of grazing animals to grassland. In this chapter some examples will be given in the studies of individual sites. In the absence of any management, slow successional changes in vegetation usually occur and these vegetational changes are also often reflected in butterfly numbers. There are numerous instances in the scheme in which increasing shade, particularly as trees and shrubs grow, is associated with slow declines of some species. It cannot be proved that these are cause and effect, but we are often confident that this is so.

The statistical test described above is objective and may seem straightforward. However, in an extreme case, there could be a local increase, relative to wider trends, even though the site was deteriorating for the species in question. This could occur, for example, in the case of a woodland species which requires open conditions; many woodland nature reserves are becoming more shaded because coppicing and clear-felling are not practised or are limited to small areas. A local population of such a species might be suffering from increasing shade, but nevertheless be increasing relative to other sites where shade may be increasing even more quickly. In other words, a departure of site values from wider trends could be caused by changes in the sites used for comparison, not necessarily the site under study. Finally, as discussed in Chapter 4, the method of calculating wider trends could itself introduce some biases. For all of these reasons, statistically significant increases or declines for a particular species are treated as provisional and requiring supporting evidence.

If a local departure from wider trends is found, several further questions, such as the following, can be asked: Do other butterflies with similar habitat requirements show similar trends at the site? Do other species with very different habitat requirements show opposite trends? Is the known management of the site consistent with the patterns of change in butterfly numbers? Has management varied in different parts of the site and are these

* i.e. based on the regression of the departure of log. site index from log. collated index over the years of recording.

differences reflected in changes in the distribution of butterflies? Thus, our assessments of the effects of management on butterflies are based on several lines of evidence and, in the end, are a matter of general judgement in addition to statistical testing.

Because we are particularly interested in effects of changes in biotopes, there is a tendency to interpret local departures of butterfly index values solely in these terms. However, at sites which are distant from the core of sites in the scheme, effects of local variations in weather are also likely to be important.

In addition to the six sites discussed in this chapter, brief summaries of the butterflies from a further 80 sites are provided by Pollard *et al.* (1986).

13.3 MONKS WOOD (in collaboration with D.D. Massen)

13.3.1 Introduction

As described earlier (Chapter 2), the methods used in the Butterfly Monitoring Scheme were developed at Monks Wood, where recording began in 1973. Monks Wood is situated 10 km northeast of Huntingdon in Cambridgeshire and covers some 157 ha. It adjoins the Institute of Terrestrial Ecology's Experimental Station, from where the Butterfly Monitoring Scheme is run.

Before 1914, the wood was managed as coppice, mainly of hazel, with standards of oak and ash (Steele and Welch, 1973). There was heavy felling during and soon after the 1914–18 war and the wood then regenerated naturally for a long period without further regular management. It was declared a National Nature Reserve in 1953.

13.3.2 Management and the butterflies

Monks Wood was once famous for its butterflies, but, unfortunately, many of the rare species for which it was known have disappeared (e.g. Thomas, 1984). The losses include the dingy skipper, brown hairstreak, purple emperor, pearl-bordered fritillary, high brown fritillary, dark green fritillary, silver-washed fritillary and marbled white. Several of these butterflies are associated with newly-cleared areas in woods and the major fellings of the 1914–18 war probably created excellent conditions for them. Eventually the regrowth of shrubs and trees led to a gradual decline of butterflies of open woodland and, by the time the wood became a nature reserve, it is likely that conditions were marginal for many of these butterflies. The marbled white survived into the early years of monitoring, but the rest were extinct by the 1960s.

The loss of these butterflies is consistent with the known changes in the condition of Monks Wood, and also with similar changes elsewhere. It remains possible that other factors were involved, but, at the very least, the woodland changes must have hastened the extinctions. Several of the rarer butterflies that have gone from the wood are now absent from all, or nearly all, of eastern England. A further change which may have affected the marbled white, and several others of the lost species, has been the loss of unimproved grassland that was once a feature of the area around the wood.

In recent years, more active management of the wood has resumed, this time with conservation as the aim. Some areas have been coppiced and rides have been managed regularly. Since the 1970s, following studies on the ecology of the black hairstreak (Thomas, 1974), areas of blackthorn scrub have been managed specifically for this butterfly. The black hairstreak is a rare butterfly which is almost entirely restricted to woodlands on heavy clay soils in the midlands. It requires blackthorn bushes in sunny, but sheltered, situations and the aim of management is to maintain a succession of small clearings with abundant blackthorn of variable age. Apart from this management of blackthorn and the limited amount of coppicing, most of the woodland compartments have remained unmanaged, and management for butterflies has been concentrated on the rides, ridesides and grass fields.

The major rides of the wood are managed to provide and maintain structural diversity. In the wide rides, the tall herb zones abutting the central, mown, grassy strip are cut on either a 2- or 4-year rotation and the cut litter removed. In any one year, the opposite sides of an individual ride are in different stages of regrowth. Between the herbs and the maturing trees of the woodland compartments is a shrub zone. This is maintained by periodic cutting which has an effect similar to coppice management, and like the latter, is also carried out in rotation. The initial programme of management of the shrub zone was from 1976 to 1978; there was then a period without such management until cutting was resumed in 1984/5.

In addition to woodland, the Monks Wood reserve includes two large fields, which were cultivated for a short period during the 1939–45 war. These fields have been managed to produce a rich grassland sward, with some scrub, and, potentially, are important butterfly areas. Each autumn the vegetation of the fields is cut in such a way to maintain a structured grassland in which selected areas of scrub are allowed to develop. Both grassland and scrub are cut intermittently, in an irregular rotation, and the cut litter removed.

13.3.3 The transect route

The route for butterfly counts (Figures 13.1 and 13.2) includes rides of various types and also runs through one of the large fields. Several of the rides of the transect route have been opened up by clearance along the edges of the adjoining woodland, as described above, during the period of monitoring. Woodland has been coppiced in compartments adjoining some sections of the route, but numerous standard trees have been left standing and the amount of sunlight penetrating to the woodland floor is much less than in most commercial coppice management. In addition to these man-made changes, section 2 (Figure 13.1) has changed substantially for other reasons. At one time this section had several large elm trees. These elms died from Dutch elm disease, mainly in the late 1970s, and their death increased the amount of light reaching the ride.

13.3.4 Changes in butterfly numbers

The fluctuations in butterfly index values of most species in Monks Wood have, for many species, been remarkably similar to fluctuations over wide areas of Britain (Pollard, 1991a; Chapter 6). At first sight, this close

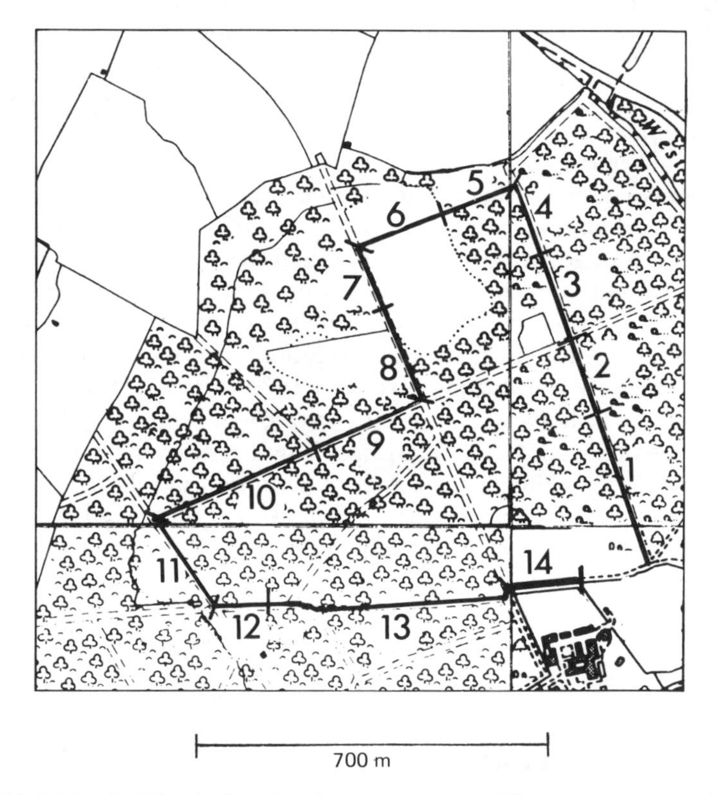

Figure 13.1 Monks Wood, showing the transect route. The route runs mostly along the major rides and through one of the two large fields (section 6). To the north of the wood is arable farmland.

Figure 13.2 Section 1 of the transect route at Monks Wood (with recorder) in 1992. This ride was one of the first to be widened (in 1972) in the programme of ride management described in the text and is now, once again, quite densely shaded.

synchrony suggests that management of the rides has had little effect on the populations of these species. However, in an earlier account of the effect of ride management on the butterflies of Monks Wood, using the monitoring data (Pollard, 1982b), clear responses of the butterflies to this management of rides were reported. The effects could be seen in both the changes of the Monks Wood index values in relation to regional trends and in changes in section index values within the transect route. How can these earlier results be reconciled with a conclusion that there has been little impact of management?

The answer seems to be that in the managed rides there were, initially, relative increases in counts of most species and declines in counts of one or two species. For example, the brimstone, orange tip, peacock and meadow brown benefited from the widening of the rides, while the ringlet found the cleared rides too sunny. Subsequently, in the later period, as the rides have once again become more shaded, these effects of management have, at least in part, been lost. In addition, unmanaged rides have become more shaded and so, over the whole period, most species remain closely correlated with the regional trends.

These effects of ride management are illustrated by results from just one of the managed rides (Figure 13.2). The increase in numbers of most species, following management, is clear and the results for the later years suggest that the benefits of management are gradually lost as the scrub at the edges of the rides regrows. The initial decline in numbers of the ringlet is also evident, but this shade-loving species seemed to benefit from the management a few years later and remains very abundant in this ride. An additional feature, clearly evident in Figure 13.2, is that the effects of widespread factors, presumably weather, are shown in the similarity of fluctuations in the Monks Wood ride to those elsewhere in eastern England. Thus even at the very local level of a single woodland ride, the tendency can be seen for the fluctuations to reflect those over much of Britain (Chapter 6).

The only locally distributed butterfly which is associated with the early stages of the woodland succession (as well as with areas maintained as grassland with scrub), and which still occurs in Monks Wood, is the grizzled skipper. This species is now scarce in eastern England. When recording began in Monks Wood in 1973 the grizzled skipper was in low numbers; it was seen mainly in the fields, although its main food plants, species of *Potentilla* and the wild strawberry, occurred widely in the wood. There was a dramatic increase in numbers in 1980 and it became abundant in several rides (Pollard *et al.*, 1986). Egg-laying females were seen in one ride (section 9) which had been recently widened (observation by E.P.). After only a short period of abundance, the grizzled skipper decreased again and it is now very rare in the wood. None were seen on transect counts from 1983 to 1991 although there have been sightings in the fields in this period. As index values have declined in East Field, as well as in the rides, we cannot assume that this species declined solely because the rides had become too shaded; we can only observe that potential breeding sites in some of the managed

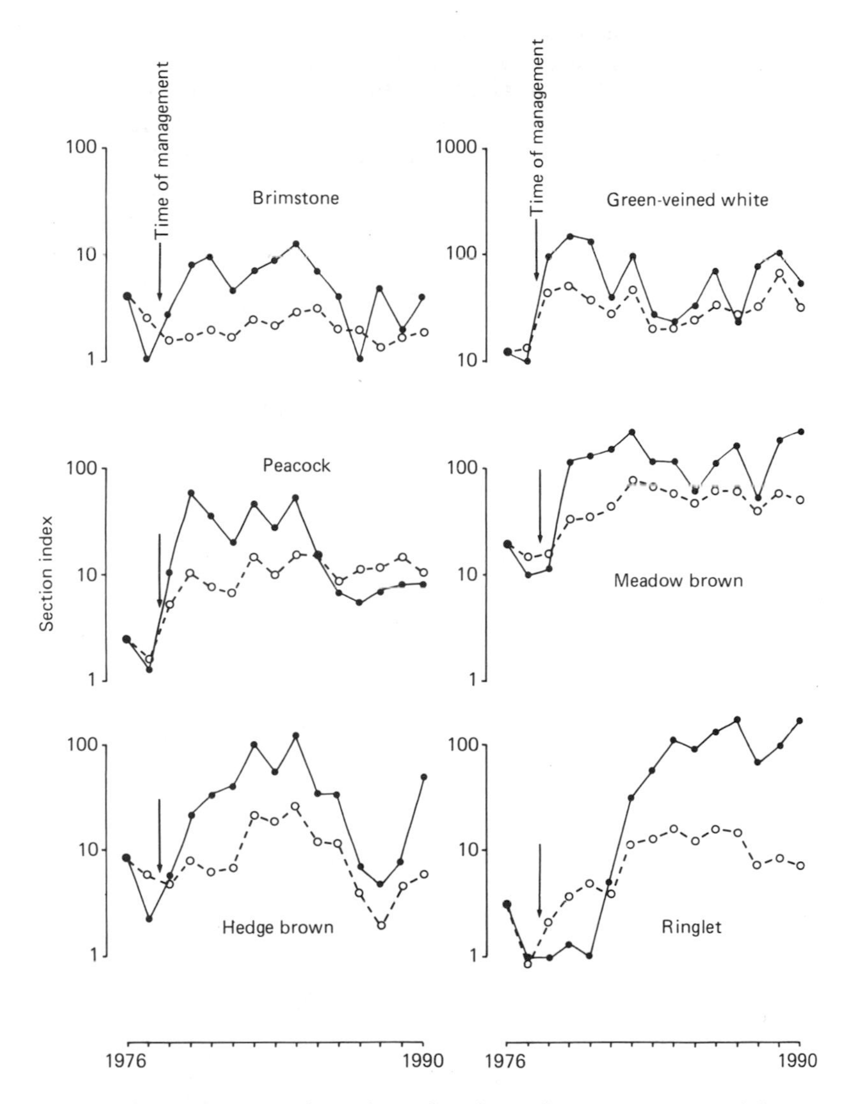

Figure 13.3 Fluctuations in index values of six butterflies in section 13 of the transect route in Monks Wood, from 1976 to 1990, compared with fluctuations at sites in eastern England. Solid line, section 13; dotted line, eastern England. Index values for the two flight-periods of the brimstone, green-veined white and peacock were summed for each year. Five-metre strips of woodland were cleared along each side of this ride in winter 1977–78, as shown by the arrow. After the clearance, most butterflies showed an early benefit, but this was largely lost as the woody plants regrew and the shade developed again. The ringlet, which prefers relatively moist and shady situations, showed an early decline after management, but later increased.

rides are now once again unsuitable. The grizzled skipper must undoubtedly be considered an endangered butterfly in Monks Wood.

Although most of the common butterflies have fluctuated in numbers in Monks Wood very much in line with wider trends, the speckled wood is a clear exception (Figure 13.4). As was described in Chapter 7, the speckled wood has colonized several sites in eastern England in recent years. At Monks Wood, it has increased enormously in abundance, compared with national trends. This butterfly is found in shadier conditions than almost any other woodland butterfly and it is possible that its increase is a response to the gradual increase in shade, in the wood as a whole, as the trees have grown taller. However, if the speckled wood's recent expansion of range in the east of England is also taken into account (Chapter 7), it seems quite possible that it has benefited from some widespread factor or factors which are at present unknown.

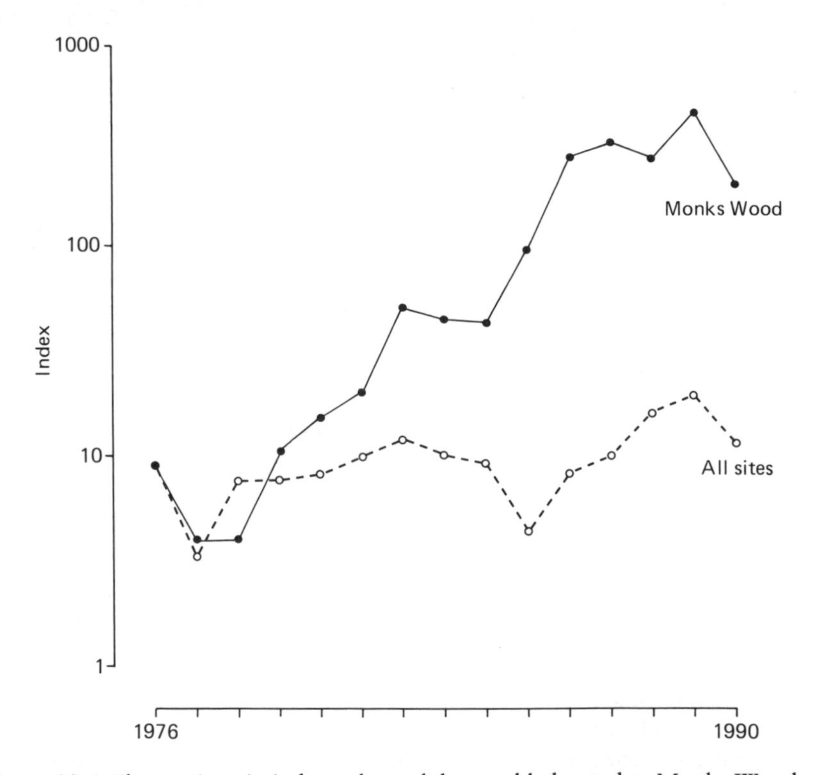

Figure 13.4 Fluctuations in index values of the speckled wood at Monks Wood, from 1976 to 1990, compared with fluctuations at all sites. The highly significant relative increase at Monks Wood ($P < 0.001$) may be partly the result of increasing shade in the woodland rides; however, the speckled wood has increased in numbers generally in eastern England and has colonized several sites (Figure 7.2).

Table 13.1 Index values of the white-letter and black hairstreaks at Monks Wood. Both of these butterflies generally fly in the tree canopy and are rarely seen on monitoring counts. The white-letter hairstreak declined after the death of elms in the late 1970s and was seen most often when many trees were dying in 1979. It survived elsewhere in the wood. The black hairstreak is a rare butterfly and Monks Wood is one of its strongholds

	1973	*74*	*75*	*76*	*77*	*78*	*79*	*80*	*81*	*82*	*83*	*84*	*85*	*86*	*87*	*88*	*89*	*90*	*91*
White-letter hairstreak	1	1	2	2	1	1	8	0	0	0	0	0	0	0	0	0	0	0	1
Black hairstreak	1	1	0	3	1	1	3	2	0	0	0	1	0	1	0	2	1	0	0

The death of elm trees, in section 2 and elsewhere in the wood away from the transect route, seems to have had its expected impact on the white-letter hairstreak (Table 13.1), which uses elm as its sole food plant. In 1979, the last of the large elms in section 2 were dying during the flight-period of the butterfly; more were seen on the counts at this time than in all the other years. It seems likely that they were behaving abnormally, perhaps dispersing from the dying elms in search of new food plants. The white-letter hairstreak was almost certainly present elsewhere in the wood throughout the 1980s (Welch, 1989), but it was not recorded again on transect counts until 1991. In Monks Wood, as in many other places, the elms have regrown from suckers. Heath *et al.* (1984) considered that mature flowering elms were required for oviposition by this species, and that young suckered trees were unsuitable. However, Emmet and Heath (1989) record that egg-laying has been observed on non-flowering elms and that larvae have completed development solely on leaves. Thus its future may not be quite as precarious as was thought when many thousands of dead elms first scarred the countryside.

The most notable surviving Monks Wood rarity, the black hairstreak, is seen only very occasionally on transect counts; like the white-letter hairstreak it seldom flies at ground level. There is a slight indication (Table 13.1) that it has become rarer in the wood over the recording period (sums of index values: 12 in the first 9 years, 5 in the next 10 years), but this variability is probably no more than might be expected from natural fluctuations. The succession of blackthorn that is maintained in the wood should ensure that it has every chance of survival.

Over many years, there have been occasional records of the wood white butterfly in Monks Wood. Most, perhaps all, of these were males and there was no evidence of breeding. There is no known population nearby, but presumably the species must have been present not too far away. In the early

1980s, there was a conspicuous increase in the abundance in the wood of one of the main food plants of the wood white, meadow vetchling. An experimental introduction of 15 female wood whites was made in 1984 and a population was established for a short period. However, they seem to have died out by 1989, perhaps because of the absence by then of low scrub along the rides and ride edges.

Two population studies of butterflies in Monks Wood are discussed elsewhere, those of the white admiral (Pollard, 1979a; Chapter 12) and the orange tip (Dempster, 1991; Chapter 14). Pollard suggested that the white admiral should continue to flourish in the wood providing that weather conditions did not deteriorate sharply. This prediction has not, at least in the short term, been realized and the white admiral has decreased in abundance in recent years. Its changes follow wider trends as closely as can be expected, given the low index values recorded throughout the period in Monks Wood, but there is certainly no indication at all of a relative increase in the wood. The reasons for its lack of success in Monks Wood are not known and this is another reminder that knowledge of the population processes of butterflies, even for those few species that have been studied in some detail, is less than perfect.

13.3.5 Conclusions

The most striking feature of the butterfly counts in Monks Wood over 18 years is that the population fluctuations of most species have remained closely similar to those of eastern England as a whole. This similarity suggests that conditions in the wood have not changed greatly in suitability for common butterflies. Equally, it is most unlikely that conditions have improved for the lost butterflies of the early stages of woodland succession.

Common butterflies such as the meadow brown and ringlet can thrive in a permanent grassy sward, such as is found in mown rides. In contrast, most of the lost butterflies of Monks Wood breed in ephemeral clearings at the stage in which woodland herbs, such as violets and primroses, are establishing themselves in bare ground. The difficulty in improving the wood for the rarer butterflies is the scale of management required. Extensive coppice management, or an alternative method of creating a succession of areas of cleared woodland, is required. As emphasized by Thomas (1991), such coppicing must include the removal of most of the standard trees; there is little benefit to butterflies in coppicing the understorey of a wood if the ground remains under the heavy shade of a high tree canopy. Recent coppicing in the wood has produced more open conditions than was the case in earlier years.

Clearance of narrow belts of trees at the edges of the rides, as has been practised to a limited extent in Monks Wood, is undoubtedly beneficial to butterflies. In addition, a recent change to a longer, 4-year, rotation for the cutting of the tall herb zone of some of the rides is likely to provide a further improvement. More frequent cutting tends to produce a rather uniform sward. If these types of management extended more widely through the wood, it is possible, although by no means certain, that populations of woodland fritillaries could be maintained in the wood. However, most of the lost species are unlikely to recolonize naturally, even if suitable conditions are created, as no populations exist nearby. Thus, any attempt to recreate conditions in Monks Wood for butterflies of early succession woodland must, to achieve its objective, include introduction of the butterflies themselves.

13.4 PICKET WOOD (in collaboration with M. Fuller)

13.4.1 Introduction

Picket Wood is on heavy clay soil in Wiltshire, just north of Westbury, and is one of three isolated woodlands in an area of pasture land and rapidly expanding urbanization. The site has almost certainly been occupied by woodland for many centuries and has an exceptionally rich ground flora with bluebells, primroses, wood anemone, violets, bugle, wild strawberry and devils-bit scabious.

Strictly, the site referred to in the monitoring scheme as Picket Wood comprises two adjoining small woods, Picket and Clanger Woods, together about 65 ha in extent. Both woods were largely cleared and replanted, mainly with conifers, between 1967 and 1973; the conifers include a range of species, with Norway spruce and European larch most abundant. Previously the woods had been planted with oak, and, over a large area, 20 m wide belts of these 50- to 60-year-old trees have been left as nurse trees for the planted conifers. Deciduous scrub is extensive at the ride edges, in some unplanted areas, and where conifers have died.

Both Picket and Clanger Woods were managed by a private forestry company, but both are now owned by the Woodland Trust, except for some 5 ha. Hence forestry is no longer the primary aim of management and the woods are used for public recreation, with wildlife conservation as a priority. Management for plants and other wildlife has included clearance of larches by the Woodland Trust in Clanger Wood, leaving stands of young deciduous trees which are mainly oak and ash. A key butterfly area, of 1.2 ha at the extreme northeast of the wood, was planted with conifers in 1980, rather later than the rest of the woodland; thus it remained relatively open after most of the rest of the wood had become heavily shaded. Recently, in 1989 and 1990, this area has been cleared of the conifers and of some planted ash and deciduous scrub.

13.4.2 The transect route

The butterfly transect includes parts of two of the three major rides of the woods and a link between these rides through dense areas of woodland along very narrow paths (Figures 13.5 and 13.6). Recent management has affected several sections of the transect route, but only the clearance of trees and shrubs in the area traversed by section 5 was completed in time to affect the butterflies in the period reported here (1981–90).

13.4.3 The butterflies

The site is one of the richest, in terms of number of butterfly species, in the monitoring scheme. Since recording began in 1981, 39 butterfly species have been seen, compared for example with 28 species in twice as long a period at Monks Wood. In part, this difference reflects the number of species that occur in the general areas of the two woods; for example, a single male chalkhill blue, a species fairly common at some sites on the nearby Salisbury Plain, was seen in Picket Wood in 1984 but certainly does

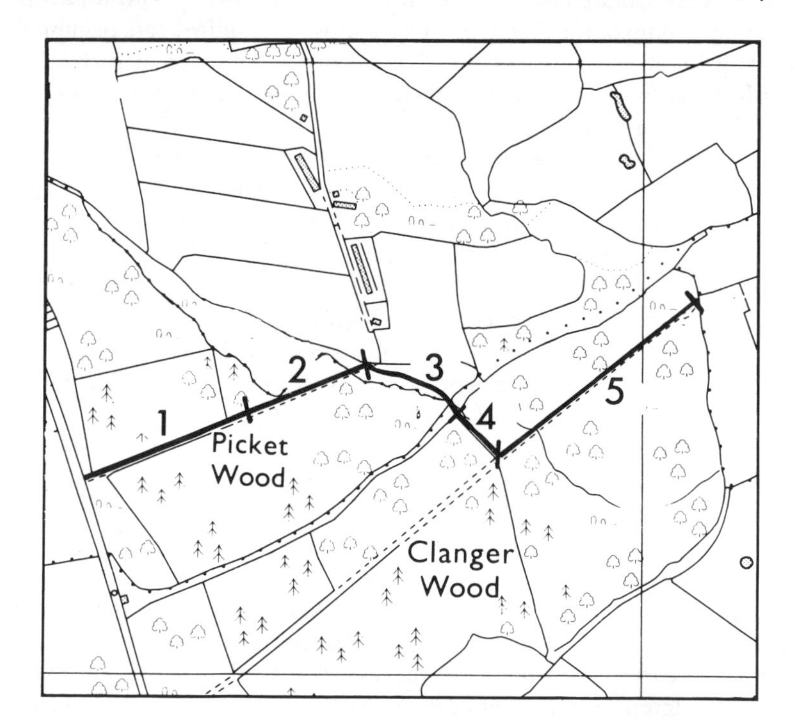

Figure 13.5 Picket Wood, Wiltshire, showing the transect route. Shrubs and trees were cleared in 1989 and 1990 in the area traversed by section 5. Scale indicated by the 1 km square grid.

Figure 13.6 Section 2 of Picket Wood in 1988. This section is along one of the three broad rides (tracks) through the wood. Most of the woodland either side of section 2 is comprised of belts of oak with conifers which were planted between them in the early 1970s. At the time of this photograph the wood was becoming too shaded for many of the rare butterflies which occur there. Extensive clearings were made to the south (right) of this ride in 1991 as part of a programme of management for the conservation of butterflies. Photograph by N. Greatorex-Davis

not breed there. Nevertheless there is a large difference in the number of species breeding in the two woods.

The breeding butterflies of Picket Wood include a number of rare and local woodland species, such as the Duke of Burgundy, white admiral, small pearl-bordered fritillary, pearl-bordered fritillary, silver-washed fritillary and the marsh fritillary, and there have been one or two records of the high brown fritillary. Most of these butterflies are now rare elsewhere in woodlands in Wiltshire and the high brown fritillary is now almost certainly extinct in the area. The rides and more open parts of the wood also support populations of butterflies more generally associated with grasslands, such as the dingy skipper, grizzled skipper, small copper, common blue, brown argus and marbled white, in addition to the full range of widespread butterflies. Considering the small and isolated character of the wood, the richness of the butterfly fauna is remarkable.

It may seem surprising that a conifer plantation contains such a large number of butterfly species, but most are dependent on the surviving ground flora of the original deciduous woodland. The clearance of most of the trees, prior to planting with conifers, undoubtedly provided good conditions for these butterflies during the early stages of the woodland succession. In many respects, the clearance provided conditions resembling the early years of a coppice cycle. Until the conifers began to cast substantial shade, their presence had little impact on the butterflies, but monitoring has shown clearly that such shading has had a major effect in recent years.

13.4.4 Population fluctuations and effects of management

Over the period since 1981, the main feature of the monitoring results has been a sharp decrease in index values of many butterflies. In some cases, such as the small copper and small heath in the examples illustrated (Figure 13.7), this trend appears to have been reversed in the last 2 years; in other cases there is little evidence of recovery.

The 1981–88 declines have generally been so steep as to make statistical tests unnecessary. There were, for example, no records of the small heath between 1985 and 1988 and it was thought that the marsh fritillary had become extinct, with no records for 5 years. There is no doubt that most of these declines were caused by a general increase in shade in the wood and it is reasonable to assume that the recent recovery has been the result of management. However, as only section 5 had been cleared of trees by the end of the recording period, it is rather surprising to find that the recovery of index values has, in many cases, occurred more widely in the wood. This is true of the grizzled skipper in the example illustrated (Figure 13.8) and also of the small copper, common blue, brown argus, meadow brown and small heath. It is possible that, in such a small wood, greatly increased breeding success in one area affects the numbers flying throughout the wood.

So far, the butterflies which seem to have benefited most from recent management of the wood are those which can perhaps be best regarded as grassland species. The butterflies which are more strictly species of open woodland, the pearl-bordered and small pearl-bordered fritillaries (Figures 13.7 and 13.8) and the Duke of Burgundy have not shown clear responses. Nevertheless, the increases in index values of several species since recent management have been encouraging, especially as this is only the start of an extensive programme of management.

13.4.5 Conclusions

Picket Wood is an outstanding example of its type, but in its rich butterfly fauna and in its decline, resembles many other conifer plantations on

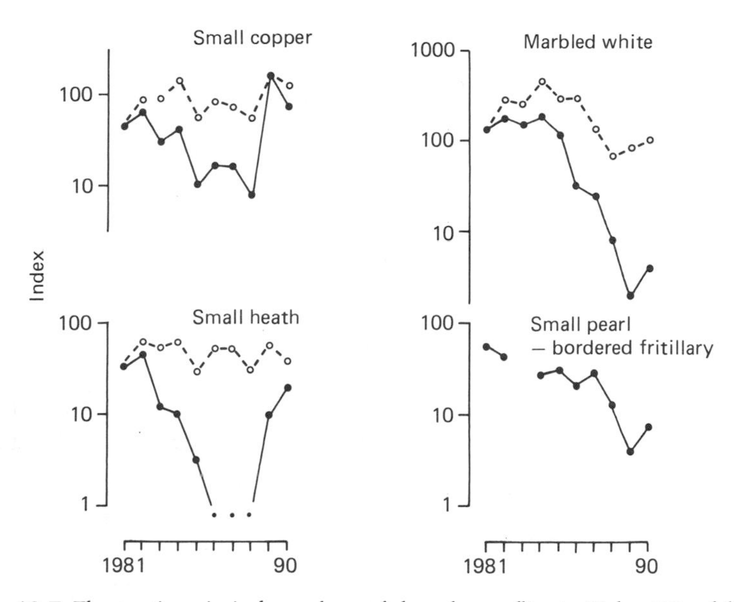

Figure 13.7 Fluctuations in index values of three butterflies in Picket Wood from 1981 to 1990 (solid line), compared with trends at all sites (dotted line). There were no records of the small heath from 1986 to 1988. No 'all-sites' indexes are available for the fourth species, the small pearl-bordered fritillary and no index value could be calculated for this species in 1983. All of these butterflies declined in abundance as conifers grew and shaded the rides. The small copper and small heath increased again following clearance of trees and shrubs in 1989 and 1990, but the other two species have not responded to management as yet.

ancient woodland sites in southern England. A few such sites are in the monitoring scheme.

The large majority of conifer plantations is, of course, much less rich and may contain only common and widespread butterflies. Nevertheless, butterflies and other insects can benefit from suitable management and the Forestry Commission have begun to manage the rides of their forests for wildlife, including butterflies. This management includes maintenance of deciduous shrubs along the edges of the rides, with periodic coppicing to maintain early succession conditions. There is no doubt that such management is beneficial to wildlife, but it has yet to be shown that management of fringe areas within a plantation can prevent the loss of some butterfly species as the conifers mature and affect the whole forest environment.

More ambitious management of conifer plantations, such as that at Picket Wood, must be regarded as a long-term policy. It is apparent that gradual clearance of the conifers, and cyclical felling of the deciduous trees which regrow, are essential if the rich butterfly fauna of some woods, now planted

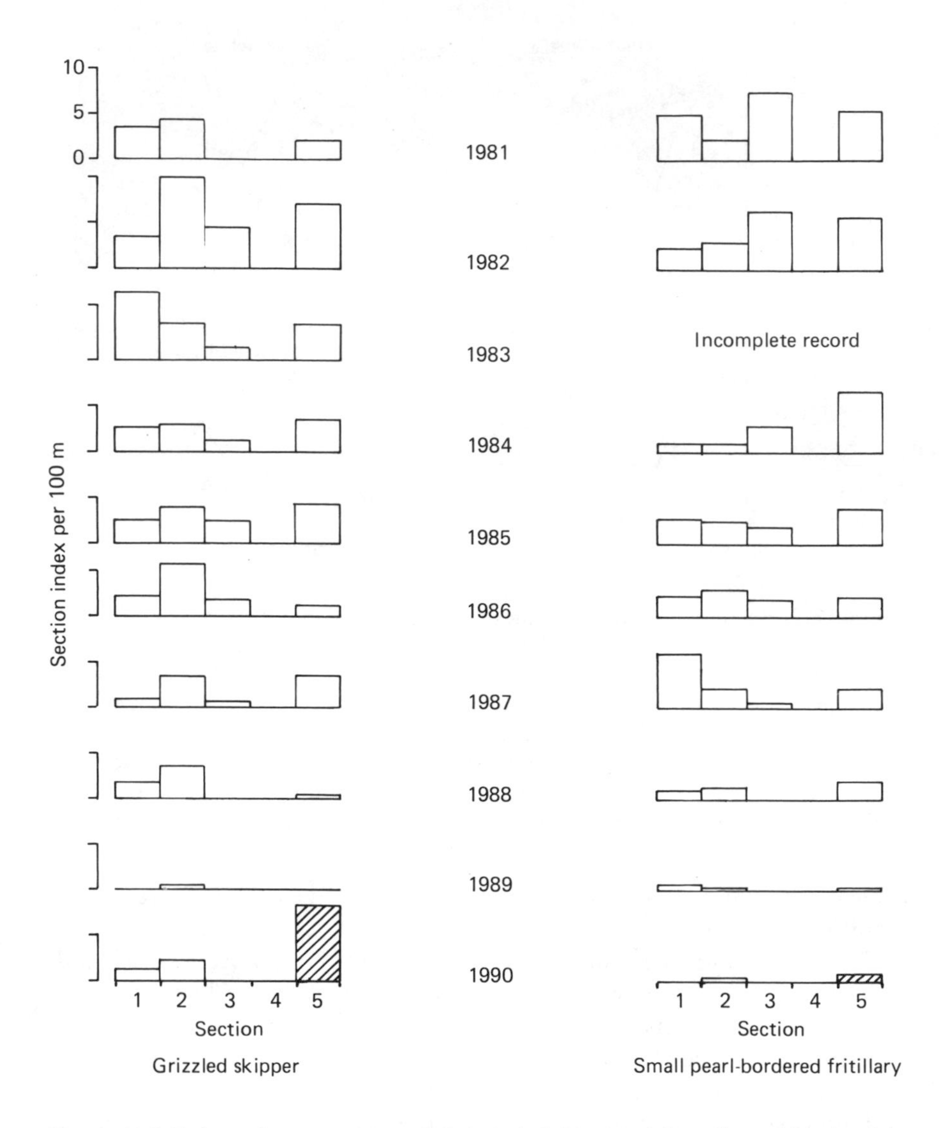

Figure 13.8 Index values per 100 m of the grizzled skipper and small pearl-bordered fritillary in the five sections of the transect at Picket Wood. The area of section 5 was partly cleared in 1989–90 and the hatched column is the count following this management. The grizzled skipper appears to have increased in abundance after clearance, but not the small pearl-bordered fritillary. However, results for further years are needed for full assessment of the effects of management.

with young conifers, is to be maintained in full or in large part. Such management cannot be achieved by a short, intensive effort, but requires long-term planning and commitment. Even where this is achieved, some butterflies may be lost because of such factors as extreme weather or the chance scarcity of a particular food plant. In a small wood, such as Picket Wood, butterfly populations are particularly vulnerable to unpredictable events. With these fairly strong reservations, the immediate prospects for the butterflies of Picket Wood seem to be good.

13.5 GAIT BARROWS (in collaboration with A.C. Aldridge and G.M. Barker)

13.5.1 Introduction

Gait Barrows, in the northwest of England, near the north Lancashire coast, has an internationally important area of limestone pavement, surrounded by deciduous woodland and meadows. Limestone pavement is a flat-topped outcrop of limestone that is mostly bare rock, but with fissures where many rare plants grow. The woodland is mainly old coppice, of ash, yew, birch and oak, with a hazel understorey. The Butterfly Monitoring Scheme transect, which was started in 1978 (G.M. Barker), follows paths and rides through the woodland, except for one section adjacent to a small, partly arable field. The main area of limestone pavement is not covered by the original transect route. When a full-time warden (A.C. Aldridge) was appointed in 1980, he set up a second transect, known as the warden's transect. This monitors those species that are common on the limestone areas but not in the woodlands, such as the dingy skipper, northern brown argus, common blue and, especially, the grayling. However, the majority of this account deals with the original woodland transect.

13.5.2 The butterflies

The area of Britain around Gait Barrows, to the south of the Lake District, is now the only real stronghold of the high brown fritillary, considered to be Britain's most endangered butterfly (Thomas and Lewington, 1991). There are also several other uncommon butterflies present, including the dingy skipper, Duke of Burgundy, northern brown argus, small pearl-bordered fritillary and pearl-bordered fritillary. The general area, which includes other monitored sites at Leighton Moss, Roudsea Wood and Wart Barrow, is exceptional for its butterflies. This may be because of a combination of the rich limestone flora and the mild local climate. The brimstone and holly blue, which are relatively common butterflies in the south of Britain, have northern outposts here, and the Scotch argus is present at its southern-most limit. The latter species was recorded on the Gait Barrows transect for the first time in 1990, but has been recorded previously elsewhere on this nature reserve.

The high brown fritillary is known to require its food plants (violets) in particular, very specific conditions (e.g. Thomas and Lewington, 1991). The female lays only in warm, sunny areas which are sheltered by a sparse growth of bracken or shrubs. The eggs are usually laid on twigs or dead leaves and do not hatch until the following spring. At Gait Barrows, a major aim of management is to provide suitable conditions for this rare butterfly.

13.5.3 The transect route and woodland management

Section 1 of the route (Figure 13.9) follows a 3 m wide track through ash, hazel and yew woodland. There is a clearing, about 100 m along, which is used as a car park. With the exception of the trees around the clearing, both sides of this section were coppiced in the winter of 1983/84, opening up what had become a very shady ride. Nine years later, the track is again

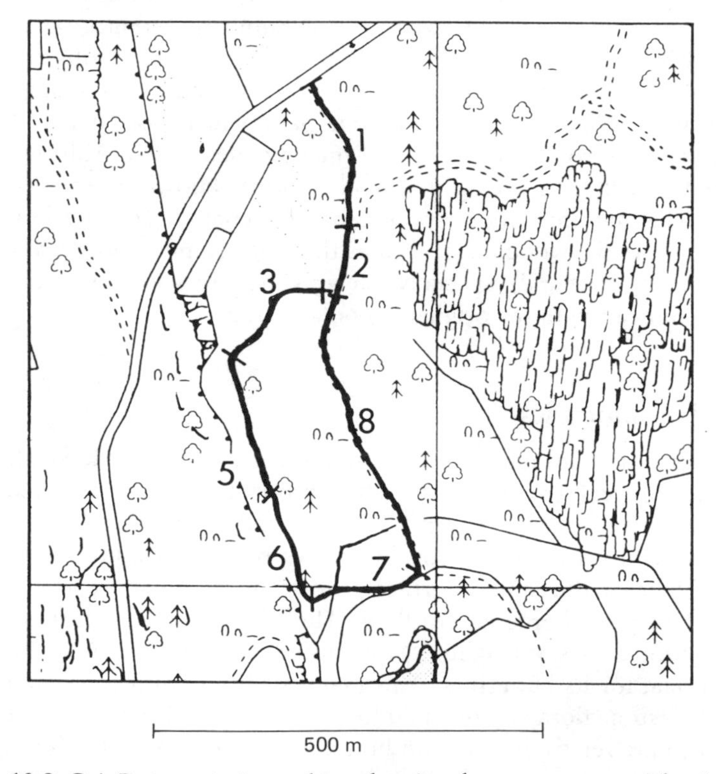

Figure 13.9 Gait Barrows in Lancashire, showing the transect route. The site includes an area of limestone pavement, to the east of the route.

becoming quite shaded, with trees on either side reaching some 5–6 m high.

Sections 3 and 5–8 (Figure 13.10) follow rides through the woodland, which is in varying stages of coppice management (Figure 13.11). Section 4, which is outside the area being managed, had to be abandoned in 1983 when it became totally impassable. It is now hard to believe that a path ever existed there. Sections 5 and 6 have a more varied ground flora than the other sections. This part of the transect has limestone outcrops, which create clearings with bare rock amongst the woodland.

The areas adjacent to all of the transect route were coppiced between 1983 and 1986 (Figure 13.11). Although some standards of oak, ash, elm and hornbeam were left in the coppiced areas, such large clearances could have led to exposure to damaging winds. This was avoided by varying the width of the coppicing, and by leaving some shrubs of spindle, hawthorn, holly and dogwood. Leaving these shrubs also ensured that the holly blue, whose larval food plants include holly, spindle and dogwood (Thomas and Lewington, 1991), was not deprived of suitable egg-laying sites in the following year.

Coppicing in the first year did not extend far from the monitored rides, but in the following year coppicing adjoining section 7 extended nearly 40 m from the ride at one place. Shelter was provided here by leaving some suitable shrubs, and by creating irregular edges to the clearings (Figure 13.11). In a similar manner, over a 3-year period, virtually the whole of the transect route changed from shady overgrown woodland to new coppice.

13.5.4 Impact of woodland management on the high brown fritillary

The coppice management at Gait Barrows has benefited the high brown fritillary, in the years following management (Figure 13.12). Although there are few comparative data from other sites for this species, these increases were coincident with coppice management and there can be no real doubt that they were direct responses. The data from the warden's transect, immediately adjacent to the woodland transect at Gait Barrows, provide a useful comparison, which further augments this conclusion. The warden's transect is naturally more open than the woodland transect, because of the limestone pavement. It has seen only limited management in the form of some clearance of birch saplings, and only a small proportion of the route has been affected by the coppicing programme. As can be seen from Figure 13.12, indexes along the warden's transect have been more stable, without the huge fluctuations seen along the woodland transect.

Comparisons of results for different areas of the transect route confirm the importance of coppicing at Gait Barrows to the high brown fritillary. Numbers generally had declined to a low level prior to the first coppicing,

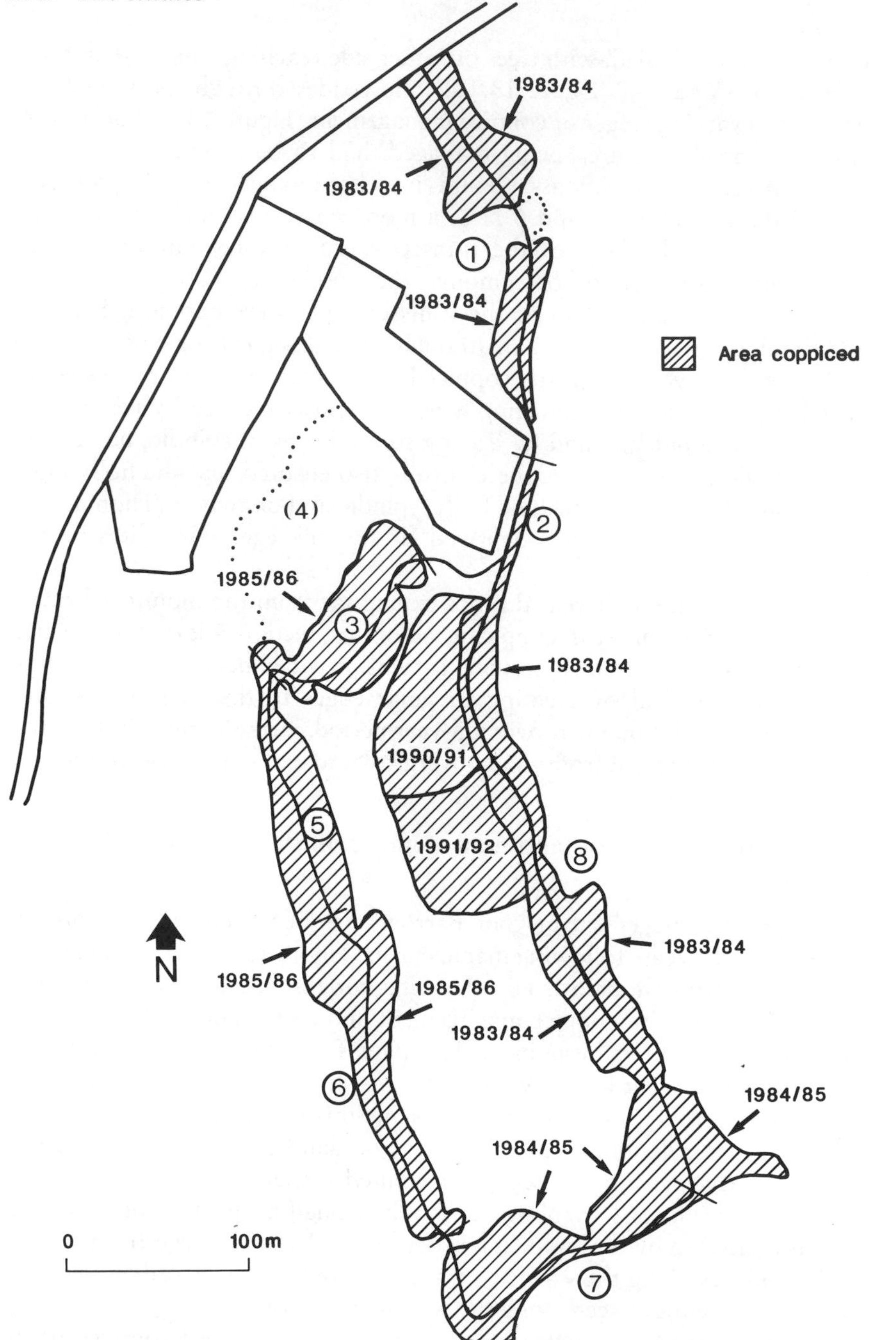

Figure 13.10 Gait Barrows, showing the transect route and areas of recently coppiced woodland (hatched), with dates of cutting.

Figure 13.11 The beginning of section 8 of the transect route at Gait Barrows showing the coppice management of the ride edges in: (a) August 1984, before management; (b) August 1985, after coppicing; (c) August 1991, coppice regrowth after 7 years. (Photographs by A.C. Aldridge, reproduced with the kind permission of English Nature.)

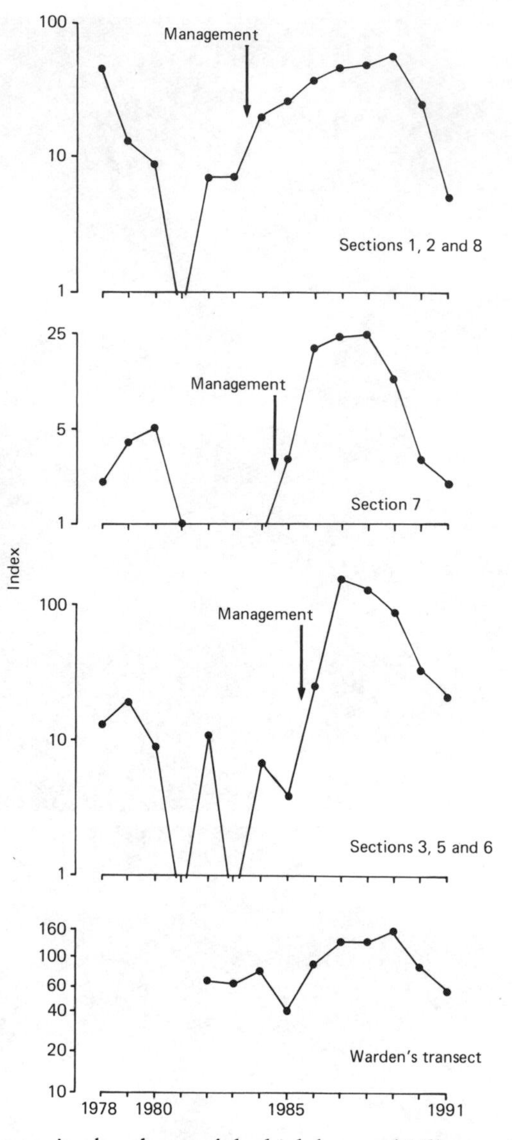

Figure 13.12 Changes in abundance of the high brown fritillary at Gait Barrows, 1978–91. The area of the main transect route was coppiced over 3 years in the mid-1980s (Figure 13.10); in the upper three graphs, the index of the main transect has been divided according to the year of management of woodland adjoining the sections of the route. In each case, the rapid response of the butterflies to coppicing is clear. In the bottom graph, index values of the warden's transect (see text) are shown. This transect route is more open, has been little influenced by coppicing, and shows no strong fluctuations.

1981 being a particularly poor year. As each area was coppiced, so numbers in these sections responded (Figure 13.12). The increases, as expected, were relatively short-lived as the shrubs grew and again began to shade the ground flora. Towards the end of the 1980s numbers were again low in the area of the transect route.

The latest coppicing, carried out during the winter of 1990/91, did not show an immediate response of a rise in the numbers of high brown fritillaries. Partly this could be due to the weather during the 1991 season, which was particularly poor during the flight-period of this species. However, another reason could be linked to a separate feature of the site. The long, linear coppicing of the mid-1980s covered a variety of terrain, while the latest management is in compact blocks. M.S. Warren (personal communication), in a study for the Nature Conservancy Council, found that the high brown fritillary required a very high local temperature for egg-laying. At Gait Barrows he discovered that the majority of eggs were laid on moss-covered limestone rocks, where the temperature could be many degrees higher than the ambient temperature. It seems probable that at Gait Barrows it is a combination of coppicing and the presence of limestone pavement that has enabled the high brown fritillary to thrive. Coppicing provides good conditions for adult flight and for some egg-laying, but the limestone pavement may be essential for the population to survive in some years.

13.5.5 Conclusions

Continued management will be needed if Gait Barrows is to maintain reasonable numbers of the high brown fritillary. With this in mind, two blocks of 0.35 ha each were selected, adjacent to section 8, for coppicing in the winters of 1990/91 and 1991/92 to give adjacent successional growth. Further coppicing is planned, taking into account the speed of regrowth of the shrub layer. The deer population in Gait Barrows browse some areas more than others, slowing down the regeneration of scrub. This means that areas frequented by the deer will need management less often than the rest of the reserve. Coppicing again, before the shrub growth has become sufficiently dense, could result in the ground flora becoming dominated by grasses, smothering the violets and bare ground that the fritillary butterflies use for egg-laying. Fortunately, the high brown fritillary population at Gait Barrows has been the subject of a detailed study, conducted by M.S. Warren, aimed at understanding this rare butterfly's needs more thoroughly. The results of this study, together with continuation of a careful management regime, should ensure a good future for the fritillaries and other butterflies of open woodland, at Gait Barrows.

13.6 WYE (In collaboration with R.V. Russell)

13.6.1 Introduction

The National Nature Reserve of Wye, on the North Downs in Kent, includes substantial areas of woodland and scrub, but it is the chalk downland which is of greatest importance for wildlife. The dominant physical feature of the reserve is the spectacular Devil's Kneading Trough, a steep-sided dry valley or 'coombe' of periglacial origin which cuts back into the scarp slope of the downs. The butterfly transect includes its steep slopes. There is a sharp change in local climate from top to bottom of the scarp, with the higher parts of the reserve often subject to strong winds which can affect the butterfly counts.

The Wye reserve, which is a more or less isolated area of rich vegetation in an otherwise intensively farmed area, is particularly notable for its orchids and other rare plants, and for its invertebrate fauna, which includes a remarkable number of rare and local species.

A major problem at Wye, over many years, has been tor grass, which can suppress other downland grasses and herbs. One of the aims of management has been to reduce the dominance of tor grass in parts of the reserve and recreate a shorter and richer downland sward. Grazing by sheep alone does not achieve this end; the sheep select the more palatable grasses and herbs in preference to the coarse tor grass; thus they can exacerbate the problem. Through the 1980s, sheep-grazing has been supplemented by grazing with cattle in late winter. Cattle graze less selectively and the tor grass is grazed, perhaps inadvertently, with other vegetation. As with virtually all other grasslands, and many other biotopes in Britain, grazing by rabbits can be considerable and this adds a further variable, and unquantified, factor.

13.6.2 The transect route

The butterfly transect route (Figure 13.13) begins at the bottom of the scarp to the west of the Kneading Trough. The first section passes through scrub, which was formerly dense but in which large clearings were created during the 1980s and are maintained by grazing with sheep in winter. Sections 2–4 climb up and along the western side of the Kneading Trough (Figure 13.14), which is open grassland with some scattered low scrub, especially on the lower slopes. Winter grazing by sheep and cattle has succeeded in creating substantial areas of short turf, where many characteristic chalk herbs are increasing in abundance, interspersed with areas where tor grass still dominates. Section 5 descends from the coombe head along one of the former meltwater channels, before crossing the head slope to join the eastern flank. On most days this section is far more sheltered than adjacent sections, resulting in higher butterfly counts.

The eastern side of the Kneading Trough (section 6; Figure 13.14) is still largely dominated by tor grass and has scattered scrub, but the final two downland sections (7 and 8), above the coombe, have been more closely grazed, by sheep and cattle, and the formerly coarse sward has been opened up. The transect route is completed by a descent towards the bottom of the scarp along an ancient sunken lane (Figure 13.14) between the woodland edge and the open downland. Here the route keeps to the top of the northern bank of the lane, adjoining grazed downland until, near the junction with section 12, dense scrub forces a descent to the bare trackway. In the last few years, livestock have been excluded from the lane and coarse grasses and scrub have increased on the banks; nevertheless, considerable areas of short turf and bare ground remain, partly due to grazing by rabbits and partly resulting from public usage. Section 12 passes between areas of scrub and woodland and is heavily shaded.

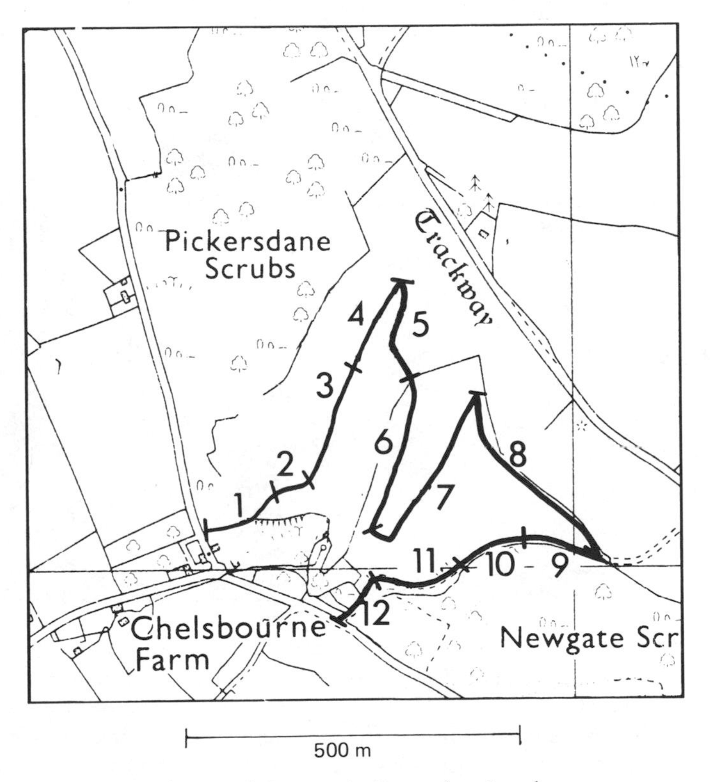

Figure 13.13 Wye on the North Downs in Kent, showing the transect route. The route ascends the scarp slope of the downs, skirts around a steep-sided dry valley and descends again along an ancient trackway.

Figure 13.14 (a) A view from the top of the scarp of the North Downs showing the Devil's Kneading Trough at Wye and the farmed landscape of Kent to the southwest beyond. The chalk grassland on the west (right of picture) slope is a mosaic composed of patches of short turf, with abundant downland herbs, and patches dominated by tor grass (sections 3 and 4 of the transect route). To the east (left), the tor grass is dominant more or less throughout (section 6 of the transect route). (b) Part of the sunken lane which comprises sections 9–12 of the transect route at Wye. Photographs taken in 1992.

13.6.3 The butterflies

As with all of the downland sites in the monitoring scheme, Wye has a rich assemblage of butterfly species, although in common with many similar sites some species have become extinct within the last 50 years. On a sunny day in August, one can expect to see the small and Essex skippers, small copper, brown argus, common blue, wall, marbled white, meadow brown and small heath. Marjoram is abundant in many areas and is a favourite nectar source for butterflies.

For some species, however, the populations are not typical of those of the other downs. The hedge brown and ringlet are both more abundant than might be expected and it seems likely that both of these species, and the ringlet in particular, benefit from the large areas of woodland and scrub that are present on the reserve and in the general neighbourhood. At several other downland sites with less woodland, such as Castle Hill and Swanage, the ringlet is absent or very rare indeed, but at Wye it is recorded even in tiny patches of scrub on otherwise open downs.

Unfortunately, Wye also differs from some of the other chalk downs represented in the monitoring scheme in the absence of two rare downland butterflies, the Adonis blue and silver-spotted skipper. Both of these species require a very short sward with a warm microclimate, although their precise requirements differ considerably (summarized by Thomas and Lewington, 1991). A third downland butterfly, the chalkhill blue, has a wider tolerance of sward heights; this butterfly colonized the area covered by the transect route at Wye during the recording period, having apparently survived in low numbers elsewhere on the reserve. It was first recorded on the transect route in 1985.

Three local, spring-flying butterflies, the dingy skipper, green hairstreak and Duke of Burgundy are recorded regularly, but generally in low numbers. Green hairstreaks have been recorded in various sections of the transect route, but the records of the dingy skipper are concentrated in the last few sections, along the sunken lane. The Duke of Burgundy has been virtually restricted to section 5 in recent years.

The comma was not seen on the Wye transect until the early 1980s, but since then has been recorded regularly. This species has been increasing in abundance and extending its range to the north. It is possible, therefore, that it has recolonized the Wye area after a period of absence. It is not a butterfly of the downs and is recorded mainly on the lower sections of the route; its main food plant in the area is likely to be nettle.

Wye is one of the closest of the monitoring sites to the continent and all of the three common migrants are recorded quite frequently here. The red admiral is seen most often in the sheltered areas at the beginning and end of the transect route. In contrast, the clouded yellow and painted lady are butterflies of open areas and, when present in the general region in any numbers, are seen frequently on the open downs. In 1983, the only year of abundance for the clouded yellow since the monitoring scheme began, 13 were seen on the Wye transect and 11 of these were on the downland sections.

13.6.4 Distribution of butterflies around the transect route

The two parts of the transect route that are particularly rich in butterflies are the winter-grazed downland (sections 2–5) and the bank of the sunken lane (sections 9–11). Distributions over the route in 1990 show clearly the importance of these areas (Figure 13.15). The summer-grazed downland sections 7 and 8 have relatively few butterflies, but here wind is a major factor and also section 8 in particular does not have a rich chalk flora. As will be discussed later, these distributions have remained essentially stable over many years. Of the species shown (Figure 13.15), it is notable that the small skipper has its highest numbers in section 6, which is dominated by rank tor grass. This butterfly lays its eggs in flowering stems of grasses, and so benefits from lack of grazing. The speckled wood also has a distinctive distribution and is most abundant in the shaded part of the sunken lane in sections 11 and 12. The heavy shade of section 12 is reflected in an abrupt decline in the general numbers of butterflies. The chalkhill blue which has recently colonized the site is already widely, if thinly distributed, although its food plant, horseshoe vetch, is sparse.

13.6.5 Recorder differences

A first look at trends in numbers at Wye indicates that many species have increased in abundance, relative to other sites. However, close inspection suggests that the Wye data provide an example of differences in recording between individuals. The Wye transect has been recorded by two main recorders, the first from 1976 to 1981, the second from 1982 to the present. Division of the data into these two periods (Figure 13.16) shows that in each period agreement with wider trends was close, but that a shift to greater abundance occurred when the recorders changed. In the examples illustrated, of the marbled white and meadow brown, we estimate that the second recorder may be recording of the order of twice as many butterflies as the first. Part of this difference could be caused by a real increase in the number of butterflies; nevertheless, the evidence for a difference between recorders is sufficiently strong for the analysis of response to change in management, through the 1980s, to be restricted to the data obtained by one recorder, from 1982 to 1990. A difference in recording between individuals is in no way a criticism of either recorder; all individuals are

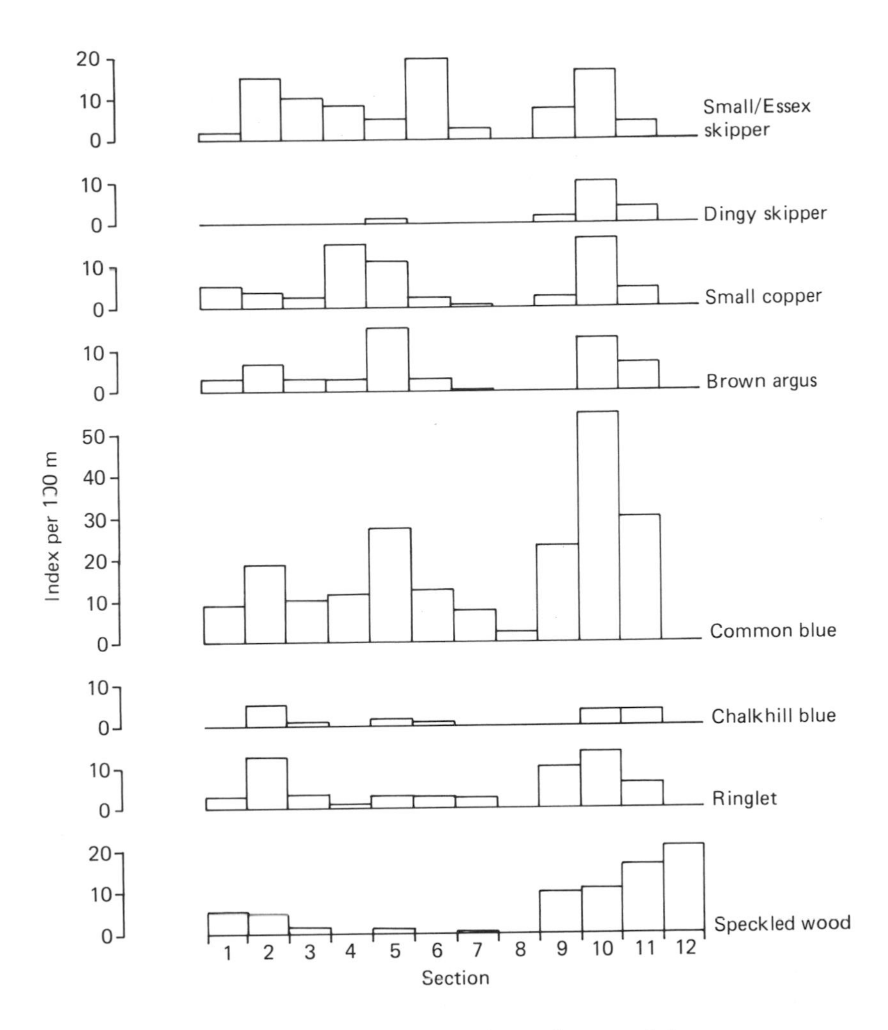

Figure 13.15 Distribution of eight species of butterfly around the transect route at Wye in 1990. Section index values per 100 m. The wide variety of biotopes at the site is reflected in the variety of distribution patterns. However, the different biotopes are often closely juxtaposed, leading to apparent overlapping of species associated with very different requirements. For example, the chalkhill blue is a butterfly of open grassland, while the speckled wood is usually found in shady woods. Nevertheless, at Wye both have been recorded in several sections. Similarly although the ringlet is, surprisingly, recorded on the downland sections, it is almost always associated there with small patches of scrub.

expected to differ to some extent in their recording and, occasionally, the differences are quite large (see Chapter 3 for a fuller discussion).

13.6.6 Population fluctuations and effects of management

Bearing in mind the possible effect of the change of recorders, population fluctuations at Wye have been close to wider trends. The two examples shown (Figure 13.16) are typical of most of the other abundant species. No species (apart from the chalkhill blue which colonized the area of the route) increased significantly relative to trends in the rest of southern England and only the hedge brown declined.

In order to examine effects of management in more detail, counts for some sections were combined for comparisons with wider trends. For example, sections 2–5 were combined to give a larger area where the dominance of tor grass had been reduced. This analysis also suggested that, so far, there has been little impact of the reduction of tor grass on butterfly numbers, with no significant departures from wider trends except for a decline in numbers of hedge browns and this species declined in all parts of the reserve.

A striking change over the 1982–90 period has been a significant relative increase in numbers of the dingy skipper in the sunken lane. Thomas and Lewington (1991) write of this species that 'Most live on warm, south-facing downland, particularly where the turf has been disturbed by grazing stock, yet is not cropped too short.' On the south-facing bank of this lane, the slope is sufficiently steep to maintain, in the absence of livestock, the

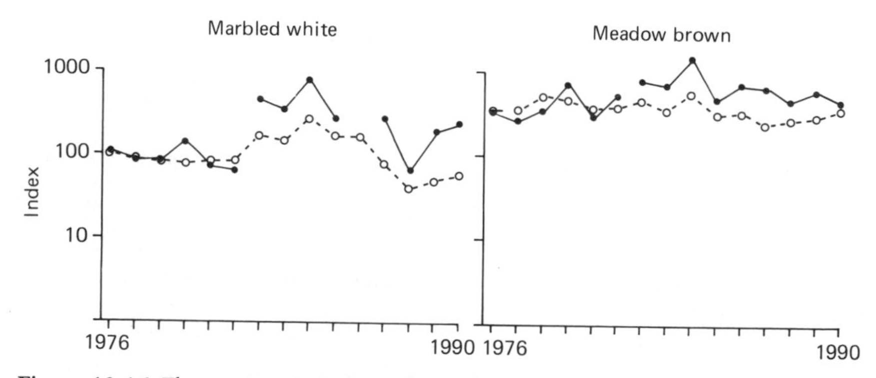

Figure 13.16 Fluctuations in index values of two butterflies at Wye, 1976–90 (solid line), compared with fluctuations in southern England (dotted line). The site was recorded from 1976 to 1981 by one recorder and from 1982 to 1990 by another. The 'step' in counts from 1981 to 1982 suggests a difference between recorders rather than a change in the abundance of butterflies. Most other butterflies showed a similar increase and there was no major change in management at this time. The difference in no way reflects on the competence of the recorders; some variation between recorders is expected and is taken into account in assessing results.

small eroded areas the dingy skipper seems to need, while the increasing shelter from patches of scrub may have benefited it.

13.6.7 Conclusions

The butterfly fauna of Wye is rich in species because of the wide range of biotopes present in the area. Monitoring shows that this rich fauna has been maintained over the recording period and indeed has been enhanced by the spread of the chalkhill blue. It is possible that its arrival in the area traversed by the transect route is a result of the opening up of the downland sward by reduction of the dominance of tor grass. However, it is disappointing that there has been no evident response to this management in terms of increases in abundance of established downland species. It is of course possible that increases have occurred, but have not been detected by our analysis for reasons outlined at the start of the chapter. Nevertheless, it seems that the recovery of the downland fauna, or at least its butterflies, following impoverishment by a long period of undergrazing, is a slow process.

13.7 TENTSMUIR POINT (in collaboration with P.K. Kinnear)

13.7.1 Introduction

Tentsmuir Point in Fife, Scotland near the mouth of the Tay estuary, includes sand flats of importance for wading birds and a large area of fixed and accreting dunes with a rich flora. Over a period of about 50 years, some parts of the dune system have grown by some 750 m; thus there are embryo slacks and dunes on the coastal fringe and mature dunes, some colonized by alder, birch and willow scrub, well inland. The sand is acidic and dune heath is a feature of the vegetation. The reserve adjoins the plantation pines of Tentsmuir Forest and colonization of dunes by seedling pines has been a long-standing problem.

13.7.2 The transect route

The butterfly transect (Figure 13.17) starts along the edge of a tall pine plantation, and continues through a damp slack of rushes, meadow-sweet and scattered sallow which is subject to flooding in winter and spring. Section 3 is through drier dune lichen and *Calluna* heath, interrupted midway by a narrow slack dominated by mature birch/willow wood. Section 4 has been dominated by invasive pine and birch (Figure 13.18), but has recently been cleared; it consists of a mosaic of low-lying grey dunes and damp slacks, characterized by cross-leaved heath and creeping willow. Sections 5 and 6 pass through one long slack, which is currently the most

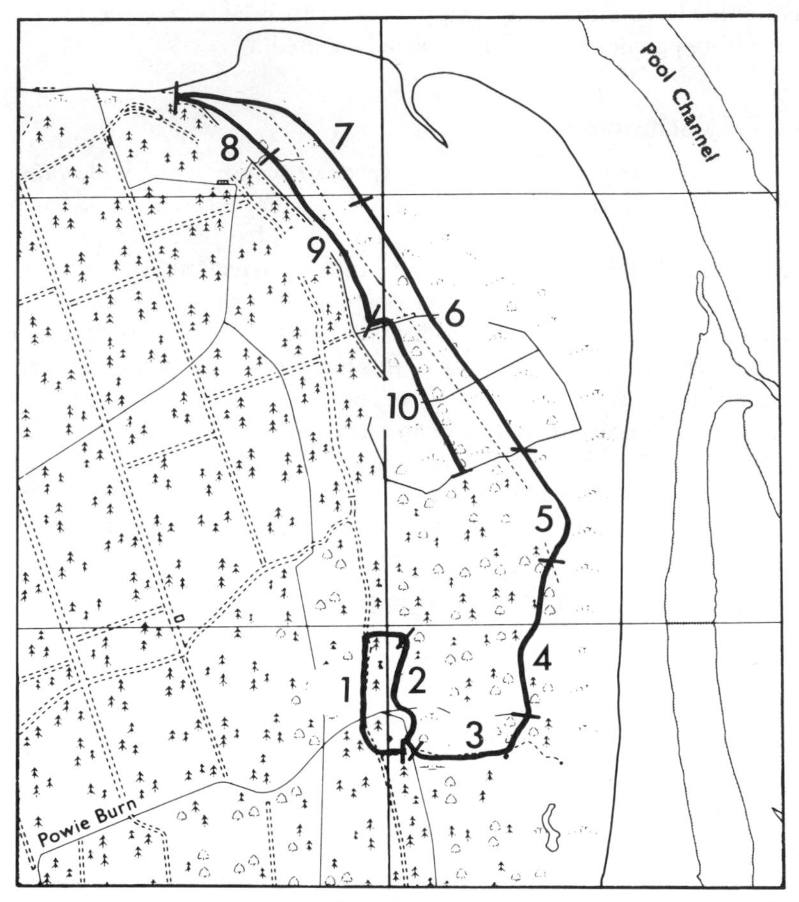

Figure 13.17 Tentsmuir Point in Fife, showing the transect route. This coastal reserve is flanked inland by the large pine plantation of Tentsmuir Forest. Some of the pine on the reserve has been cleared since the date of this map.

coastal. This slack was formerly dominated by rushes and low-growing herbs such as grass of Parnassus, but creeping willow and, in places, birch are invading. Section 7 is through marram grass with some rose-bay willow herb and the last three sections return to older dunes, dominated by grasses and sand sedge, with varying amounts of lichen and *Calluna* heath. Section 10 is both shaded and sheltered by mature birch/alder woodland along the east side.

13.7.3 The butterflies

The Tentsmuir butterflies include perhaps the largest population of the grayling in the monitoring scheme (the index value has been the largest in

Figure 13.18 Section 4 of the transect route at Tentsmuir Point in 1989. The area is a mosaic of dunes and slacks and at this time was dominated by invasive pine and birch, which have since been cleared. The pines of Tentsmuir Forest can be seen in the background.

almost every year). Other species of special interest include the green hairstreak and dark green fritillary. Both of these butterflies occur on a wide range of biotopes elsewhere in Britain, from chalk downs with some scrub to moorland, but both have suffered from the increasing intensity of land use, especially in the south. If these losses continue, northern sites, such as Tentsmuir, will become increasingly important for their conservation.

Of the other Tentsmuir butterflies, the green-veined white, small copper, meadow brown, small heath and ringlet are all resident species which are abundant in some years. Apart from the neighbouring reserve, Morton Lochs, this is the northernmost site for the ringlet in the monitoring scheme. In the south of England the ringlet is associated with damp woodland rides and other shady areas, but at Tentsmuir, as elsewhere in Scotland, it also occurs in rather more exposed situations, even flying with the grayling in the open areas of section 5.

In Chapter 3, an analysis of data for the common blue at Tentsmuir, where it has a single generation each year, suggested that the flight of butterflies was much reduced in strong winds. Other butterflies are likely to be similarly affected, both by the wind directly and by the low coastal temperature. For example, at Morton Lochs, a monitored site just inland, the first generation of the green-veined white regularly emerges about a week earlier than at Tentsmuir and also has a longer flight-period (unpublished data).

Thomson (1980), in his excellent book on Scottish butterflies, states that the small tortoiseshell has only one generation each year in all parts of Scotland and that any individuals seen in late August and September have migrated from southern Britain. At Tentsmuir there were many such late-summer individuals in the warm season of 1984 and at Morton Lochs in 1982 and 1984. In the absence of records of larvae, we cannot be absolutely certain, but suspect that these late records were from local breeding.

13.7.4 Population fluctuations

In common with northern sites more generally, fluctuations in butterfly numbers have been larger at Tentsmuir Point than further south in Britain (Figure 13.19). Other butterflies, such as the green hairstreak, common blue, small tortoiseshell and small heath have shown even greater variability here than in the examples illustrated, with the index of the common blue, for example, ranging from 4 in 1986 to 156 in 1989. However, population fluctuations of Tentsmuir butterflies have been generally less extreme than at Scottish sites further north.

Tentsmuir is remote from the main nucleus of monitoring scheme sites and fluctuations in numbers show less synchrony with trends at all sites than is usually the case. Nevertheless, in all of the examples illustrated in Figure 13.19, the Tentsmuir index values are positively correlated with the 'all-sites' index; in the case of the dark green fritillary the correlation is significant.* We interpret this generally weak synchrony between butterflies at Tentsmuir and those in Britain more generally to indicate that the climate at Tentsmuir is more subject to local variation than at most of the sites in the scheme, but nevertheless is influenced by the broad pattern of annual variation in weather experienced over Britain. As yet, no analysis has been made to examine the effects of local weather on the butterflies of Tentsmuir.

Of the migratory butterflies, the red admiral has been recorded in most years, but the painted lady in only five of the 13 years. The large majority (70%) of the total records of the painted lady at Tentsmuir, over all years, was during the great immigration along the east coast of Britain in 1980 (see Chapter 9). This migration was too late in the season for the butterflies to produce a new generation before the onset of winter.

* $r = 0.64$, $P < 0.05$.

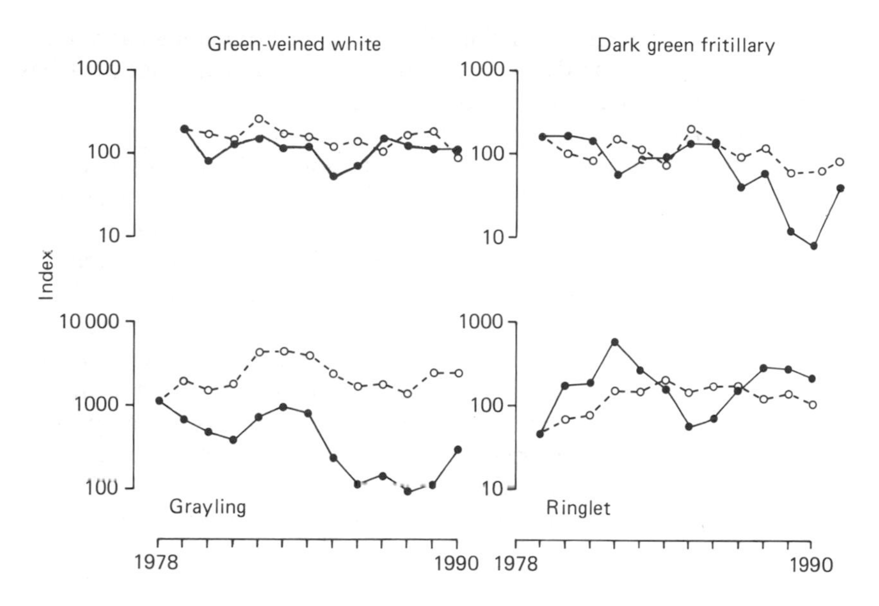

Figure 13.19 Fluctuations in index values of four butterflies at Tentsmuir Point, 1978 (or 1979)–90 (solid line), compared with fluctuations at all sites (dotted line). In 1978 only the latter part of the season was recorded. In the case of the green-veined white, the index values for the two generations are summed. Both the dark green fritillary and the grayling have declined significantly ($P < 0.001$) at Tentsmuir, relative to the all-sites indexes.

13.7.5 Distribution around the transect route

The butterflies at Tentsmuir fall into two main groups as regards their local distributions. The first, smaller group, consisting of the green-veined white, green hairstreak and ringlet, is recorded most frequently in the more sheltered, or moister, first few sections of the route. It is notable that most sightings of the green hairstreak are near the only patch of bilberry, a major food plant, that occurs on the reserve. The second group includes the small copper, dark green fritillary, grayling, meadow brown and small heath, and is associated particularly with the more open dunes of the last few sections. The common blue falls into neither of these categories, with most records in sections 4–6, where there is an abundance of its main food plant, birdsfoot trefoil.

13.7.6 Effects of local vegetation changes and management

Although changes in vegetation along the transect route have not been specifically monitored at Tentsmuir, reduction in grazing by rabbits is thought to have led to a decrease in the number of small patches of bare ground and led to some impoverishment of the dune flora along parts of the transect route. Scrub has developed in most slacks and management has included scrub clearance as well as removal of pine.

Analysis of trends in index values for each section of the transect route indicates only minor changes in the distribution of butterflies over the recording period. Most butterflies have declined in abundance, relative to wider trends shown at other sites, along sections 7–10. Here the recorder (P.K.K.) has noted a decline in grazing by rabbits, with a likely reduction in the number of nectar sources, such as thyme, and of some larval food plants, such as the violets used by the dark green fritillary, and sheep's sorrel used by the small copper. The grayling (Figure 13.20) has declined significantly in all of these sections; it is a species known to be associated with fine grasses growing amongst bare earth, 'where the sun bakes the ground' (Thomas, 1986). A decline in grayling numbers would clearly be expected as a result of a reduction in rabbit grazing.

The transect route is not close to the areas of the reserve where most trees and scrub have been cleared. Section 4 is the section most likely to be affected by the general recovery of vegetation following such clearance and we find that this is the only section of the transect route to show minor, relative increases of several species. In this section the small copper, dark green fritillary, meadow brown and grayling have all increased slightly, although not significantly, relative to wider trends.

Although the main transect route has not been greatly affected by the removal of trees and scrub, an additional area has been recorded to monitor

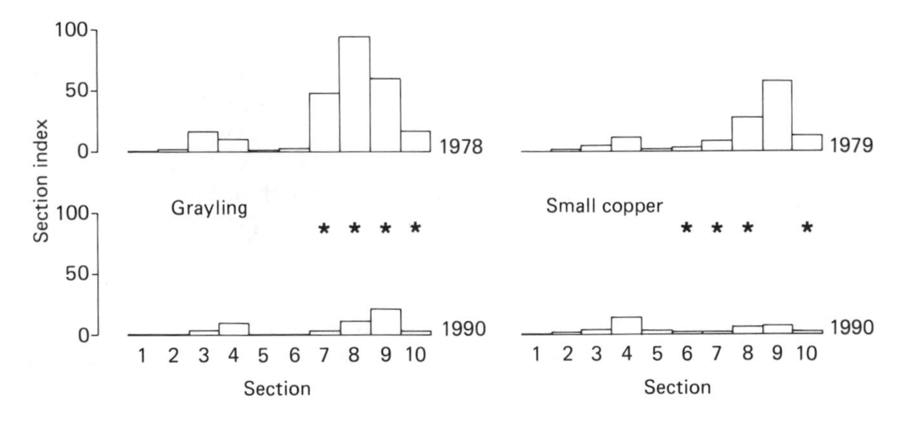

Figure 13.20 Section index values per 100 m of the grayling in 1978 and 1990 and of the small copper in 1979 and 1990 at Tentsmuir Point.* A significant decline ($P < 0.05$), relative to all sites trends, over the recording period. Numbers were high in the last few sections in the late 1970s, but have declined in recent years, perhaps because of changes in vegetation following a reduction in grazing by rabbits.

changes in butterfly numbers following the felling of pine trees. Prior to 1985 this area, consisting of low-lying dunes and slacks (which were rarely flooded), was densely covered with seeded Scots pine, with some birch and pockets of willow and alder. The few butterflies present were in the remaining more open areas. Felling of pine began in early 1985 and the last mature birch were removed by 1989. The damper areas have been colonized by willow, cross-leaved heath, valerian and ragged robin, and heath vegetation is returning to the dunes.

The response of butterflies to this management has been rapid and is shown clearly by the counts on the additional transect route (Table 13.2). It is not surprising that the felling of trees enables more butterflies to fly in an area, but the colonization by several species that are considered to be sedentary and to fly mainly within their breeding areas is encouraging. These butterflies include the grayling and small copper, which have been shown to be declining in unmanaged areas of the reserve.

13.7.7 Conclusions

The gradual maturation of a dune system must be expected to result in changes in vegetation and in the abundance of butterfly species. At Tentsmuir Point, this natural progression has been deflected by colonization by pine trees from the nearby commercial plantation and perhaps speeded by a reduction in grazing by rabbits. A decline in abundance of some of the

Table 13.2 Counts of some butterfly species along the additional transect sections (a) and (b) at Tentsmuir Point (see text). Section (a) is pine woodland; section (b) was pine woodland, but was cleared in 1985 and 1986. The final clearance also opened up the end of section (a) to some extent

	1985	*1986*	*1987*	*1988*	*1989*	*1990*
(a) Pine woodland						
Small copper	1	1	1	1	1	1
Common blue	0	0	0	0	4	1
Grayling	1	2	3	10	1	1
Meadow brown	1	1	1	5	3	1
Ringlet	0	0	0	0	0	5
(b) Cleared woodland						
Small copper	8	8	4	15	6	11
Common blue	0	0	3	6	16	2
Grayling	18	26	37	22	25	36
Meadow brown	1	2	11	25	13	11
Ringlet	0	1	6	17	23	6

butterflies of the open dunes has, to some extent, been offset by clearance of the pine and other woody plants. However, in the longer term, the formation of new dunes, and the establishment of vegetation on them, is creating new areas suitable for the grayling and other butterflies of open ground.

Monitoring such a rapidly-changing site will in the long term present problems. Inevitably, the existing transect route, which was adopted in 1978, will change further as the vegetation continues to mature; more importantly, newly vegetated dunes will not be included and the route will become less representative of the site as a whole. In time it may be necessary to adopt an additional transect route to cater for these changes.

13.8 ST OSYTH (In collaroration with R.W. Arthur and B.C. Manning)

13.8.1 Introduction

The dune vegetation of Tentsmuir is as natural as any vegetation in Britain, and the ancient downland of Wye is natural in the sense that the grasses and herbs have not been sown or planted by humans. However, much of the vegetation of the restored landfill site of Martin's Farm at St Osyth near Clacton-on-Sea in Essex has been sown recently and the species composition determined initially by humans; thus it provides a sharp contrast with other sites described in this chapter.

In 1982, Essex County Council proposed the creation of a country park from the waste disposal site at Martin's Farm. For this purpose, the establishment of grassland and areas of trees was planned, although the area will not be opened to the public until waste disposal is completed on a nearby site. The Institute of Terrestrial Ecology was commissioned to create an attractive grassland which would attract a range of butterflies. This account is based in large part on accounts of this project by Davis (1989) and Davis and Coppeard (1990), in which the development of vegetation and establishment of butterflies at St Osyth are described.

13.8.2 Establishment of vegetation

The area of the site is about 24 ha, with the western half sloping down to tidal salt-marshes. In autumn 1982, soil (0.6 m depth) was brought in to complete the covering of compacted waste in the northwest part of the site; the oldest area to the southeast was already covered with a dense growth of couch grass, which colonized the area naturally. The establishment of vegetation was studied on 12 replicated plots (about 0.3 ha each), but these will not be discussed in detail here (Davis, 1989).

Two grass mixtures, containing 10 species of grasses and 21 herbs, were used in some plots. The herbs included several which were chosen as larval food plants for butterflies: kidney vetch, lady's smock, birdsfoot trefoil, black medick and common sorrel. The grasses included species known to be food plants of satyrid and hesperid butterflies (browns and skippers). Other plots were either untreated controls, or were not sown and were cultivated occasionally. All plots except the controls were cultivated in August 1983 and a seed-bed prepared. Apart from the application of a little fertilizer to promote growth in the poor soil, and the harrowing of the cultivated plots, there was no other management.

13.8.3 Monitoring vegetation development and butterflies

Vegetation structure and composition were recorded and a butterfly transect was established (Figure 13.21). The route shown has nine main sections, though these were subdivided to provide separate counts for each plot within the experimental area (Davis, 1989). Section 1 of the simplified route passed through the tall couch grass area; in an attempt to prevent the spread of couch, this area was cultivated in spring 1987, sown with an agricultural grass/clover mixture, and cut and baled for hay.

The establishment of the sward was not without problems. The shallow soil, overlying compacted spoil, was subject to drought in summer and waterlogging in winter. All 10 of the grass species eventually became established, but abundance varied greatly. Lady's smock and greater knapweed (a good nectar source for butterflies) failed completely. Hedge garlic was sown later as a replacement for lady's smock as a food plant for the orange tip and green-veined white, but this too largely failed. In contrast, two food plants of the common blue butterfly, birdsfoot trefoil and black medick, flourished.

The experimental plots were very varied in species composition and structure. Many plant species colonized naturally and a total of over 140 species was recorded eventually. Several good nectar sources for butterflies, such as thistles and teasel, were prominent amongst these colonizers and a large patch of nettles also developed where sewage sludge had been used.

Between 1983 and 1991, 21 butterfly species were recorded on the transect route. The most detailed information is available for the period of the experimental study, from 1983 to 1987, but the butterfly transect has been continued (Table 13.3). Davis (1989) suggests that the sequence of establishment of butterflies has been meadow brown, followed by small/ Essex skipper and small heath, all before 1983 when the counts began; these were followed by the common blue and the hedge brown, probably in 1983.

Peacock caterpillars were seen on the nettle patch in 1983 and those of the small tortoiseshell in 1987. Both of these butterflies are wide-ranging species which do not form local populations, but lay eggs opportunistically

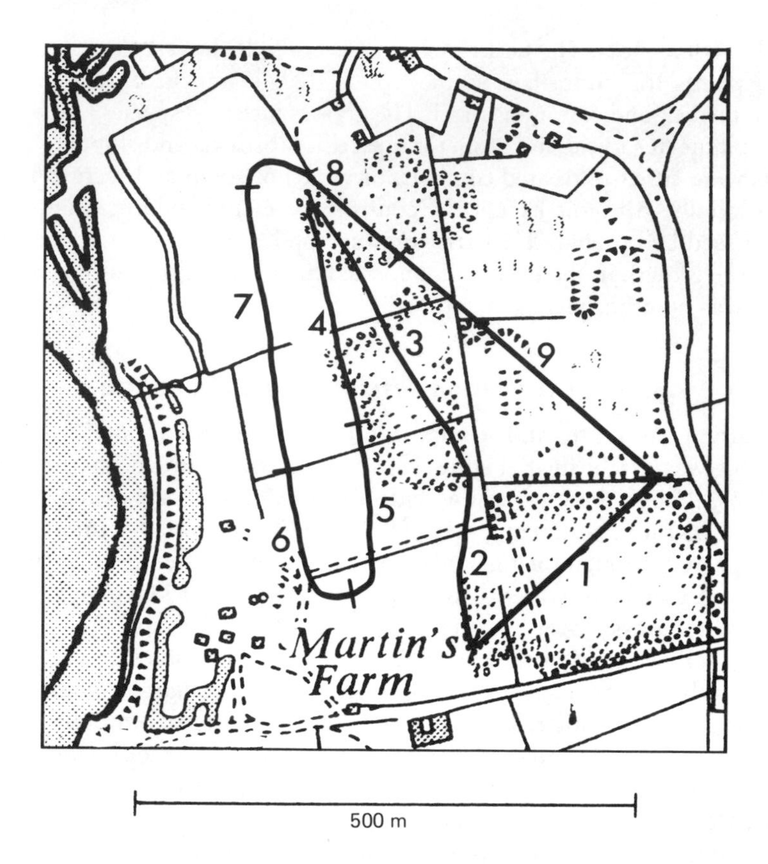

Figure 13.21 St Osyth landfill site in Essex, showing the transect route. Soil was transported to the site to cover compacted waste; in some areas (sections 1–3) a grass/clover mixture was sown, and in others (sections 4–7), a rich mixture of grasses and herbs.

wherever they find suitable food plants. The green-veined white was seen first in 1987, but it is not known if it bred on the site. Similarly, several other species have been recorded in low numbers, but local breeding has not been confirmed. Thus, in total the site supports breeding populations of six or seven species and provides a breeding site for others which do not form local populations. In addition, butterflies such as the small white, red admiral and orange tip are attracted to the area by the variety of flowers.

Several butterflies (the common blue, small tortoiseshell, small heath and meadow brown) increased dramatically in numbers in the early years of the establishment of the grassland (Table 13.3). The period of increase of the common blue was associated with the establishment and spread of birdsfoot trefoil; it was recorded most frequently in the plots sown with this species.

Table 13.3 Index values of seven butterflies at the St Osyth landfill site in Essex, 1983–91. The results show, in general, the increase in butterfly numbers as floristically rich vegetation was established at the site. A dash indicates that data were insufficient for the calculation of an index. The two lines of data for the small copper and common blue are for the two flight-periods

	1983	*1984*	*1985*	*1986*	*1987*	*1988*	*1989*	*1990*	*1991*
Small/Essex skipper	76	38	32	71	236	386	560	84	1126
Small copper	0	1	0	2	–	–	0	0	–
	–	1	1	1	1	3	7	2	16
Common blue	5	19	33	60	313	–	6	–	–
	–	28	148	102	195	–	5	2	126
Small tortoiseshell	4	13	48	68	382	48	28	21	–
Hedge brown	5	5	13	21	48	41	126	55	173
Meadow brown	144	98	283	214	960	821	819	175	518
Small heath	55	140	514	635	623	–	299	175	–

Interestingly however, no larvae were found on this plant which was a tall-growing agricultural cultivar as seed of the native strain was not available at the time. All larvae of the common blue were found on low-growing black medick. Numbers fell again sharply in 1988–90 when fertilizers encouraged dense growth of grasses around the birdsfoot trefoil and black medick. Drought may have reduced the abundance of these food plants but by opening up the sward the drought may also have helped to recreate favourable conditions, leading to a recovery of the common blue in 1991.

Davis (1989) thought that the grass-feeding species in particular would continue to flourish and, so far, this has proved to be the case. There has been a gradual increase in numbers of the hedge brown and the small/Essex skippers reached a new peak in 1991. However, the absence of any shelter on the site seems to have deterred colonization by such species as the large skipper. The Essex skipper, as its name suggests, was first identified as a British species in Essex in 1888; appropriately it generally outnumbers the small skipper at St Osyth.

The counts for the individual experimental plots did not show clear differences between treatments for most butterfly species. This was partly because the plots were small (approximately 60 m × 50 m), so that butterflies flew readily between them, and partly because plots of the same treatment developed very different vegetation. Thus the monitoring scheme data proved particularly useful in providing data from other sites which could be used for comparison with the St Osyth counts.

As part of the research programme, two butterfly species, the marbled white and grayling, were introduced to the site. The conditions were

Figure 13.22 Section 6 of the transect route at the landfill site of St Osyth, with Flag Creek in the background. In this part of the route various mixtures of grasses and wildflowers were sown in 1983, to create conditions suitable for butterflies and other wildlife. The photograph was taken in 1992.

considered suitable, but existing populations were too distant for natural colonization to be likely. Unfortunately, the counts suggest that both of these introductions have failed. It might have been expected that the grayling in particular could cope with, and indeed thrive in, the summer droughts to which the site is prone. However, there have been no detailed studies of the habitat requirements of this butterfly and it is possible that important requirements for the species were not present.

13.8.4 Comparisons with wider trends in butterfly numbers

It is clear from the monitoring results (Table 13.3) that butterfly numbers at St Osyth have increased dramatically. That this is a local feature is confirmed by comparisons with fluctuations in eastern England for two of the more abundant St Osyth butterflies (Figure 13.23). Both have increased relative to wider trends, and the same is true of several other species. Thus monitoring has shown, with very little room for doubt, that the increases are related to the establishment of the rich sward.

In a number of comparisons of this type (Figure 13.23) in this and other chapters, attention has been drawn to synchrony between local populations and wider fluctuations, based on all of the sites in a region or in Britain. In the case of St Osyth (Figure 13.23) there is no such synchrony; not only are

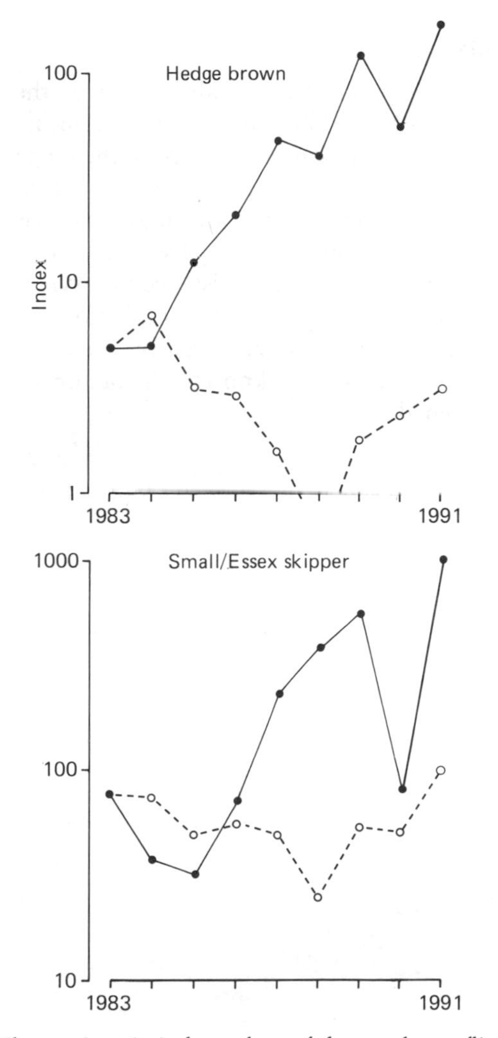

Figure 13.23 Fluctuations in index values of the two butterflies at St Osyth (solid line), compared with fluctuations in eastern England (dotted line). Unlike almost all other sites, there is little or no similarity between site and regional fluctuations. The vegetaion has developed rapidly and the butterflies have responded to these changes; any effects of weather on butterfly numbers have been overwhelmed by these vegetational changes.

the longer-term trends sharply different from those in eastern England, but there is no similarity in changes from one year to the next. St Osyth has differed from other sites in the very rapid changes in its vegetation, and these rapid changes are also reflected in changes in the local butterfly populations.

13.8.5 Conclusions

This study illustrates the potential, and also some of the problems, of establishing new biotopes for wildlife. In the short term, it seems unlikely that this site will support a wide range of species; in the longer term, if more trees and shrubs can be established and if the organic content of the soil increases, more of the commoner butterflies may colonize. It is only by studies of this type that the conservation potential of landfill and other reclaimed areas can be assessed and the existence of a synoptic monitoring scheme has helped considerably in the assessment. Unfortunately, studies at the St Osyth site have been partly interrupted by a recent resumption of tipping. At the time of writing it is not known whether or not its future as a Country Park is threatened.

14

Population ecology

14.1 INTRODUCTION

Population ecology is the study of characteristic abundance of populations of plants or animals, their spatial and temporal fluctuations and the factors which determine all of these. In extreme cases, population declines may lead to extinction, while dispersal of individuals may lead to the foundation of new populations elsewhere; thus population ecology leads into the study of changes of range and the biogeography of species.

In a general way, the subject of this book as a whole is the acquisition and interpretation of information on the population ecology of butterflies. For example, in Chapter 6 the variability of populations is discussed, in Chapter 8 the effects of weather on population fluctuations are described and in Chapter 13 the effects of management on butterfly populations are described. In this chapter we bring these, and other, strands together.

The British butterflies include some species with individuals which may fly across continents and others in which a discrete population may occupy a few square metres of land for many generations. We are most concerned here with butterflies that occur in more or less discrete populations. Clearly, a general framework should, by definition, include all species; that described here may do so, but the great mobility of some species adds an extra dimension which is, for the present, not encompassed.

This chapter is divided into three sections; first, relevant features of data from the monitoring scheme are summarized; second, the discussion is broadened to include evidence from a wider range of studies on butterflies and other insects; finally, a framework for interpreting the population ecology of butterflies is outlined.

14.2 EVIDENCE FROM THE MONITORING SCHEME

The special contribution of the Butterfly Monitoring Scheme to an understanding of the population ecology of butterflies lies in its synoptic nature. Although the data lack any of the detail of individual population studies and do not give direct information on the factors which determine the size of populations and cause them to fluctuate, the broad view they provide has not been available before.

As the results of the monitoring scheme have accumulated over the years, a number of general features have become clear. The most striking of these are: (1) the pattern of fluctuations in index values of individual butterfly species from year to year is often very similar from site to site; a corollary of this is that: (2) the relative size of index values from site to site often remains very similar over many years; similarly, the relative abundance from section to section of transect routes (the local distribution) is often very similar from year to year. (3) In spite of this general synchrony and stability, index values of some species at some sites depart from wider trends; the time-scale for these departures may be gradual or, in a few cases, rapid. Similarly in these instances, local distributions may change considerably over time. The nature of these exceptions helps to provide an understanding of the reasons for the more usual synchrony and stability.

14.2.1 Synchrony of population fluctuations

In general, the fluctuations in index values of individual butterfly species have been similar over large areas of the country (Chapter 6; Pollard, 1991a). Annual fluctuations have been shown to be correlated with variation in particular weather conditions (Chapter 8; Pollard, 1988) and weather undoubtedly plays a major role in the synchrony of index values.

Some butterfly species may be so wide ranging that a large area of Britain may be regarded as the area occupied by a population unit. This is probably true, for example, of the large white, the peacock and small tortoiseshell; in such instances synchrony might be expected as many index values will be, effectively, samples from a single population. Nevertheless, it is in this group that regional differences in population fluctuations have been most pronounced, for example in the declines of the brimstone, peacock and small tortoiseshell in the north of Britain in the mid-1980s.

Many common, and most rare, British butterflies are considered to occur in populations from which there is little immigration and emigration (Appendix A). Although it is perhaps doubtful whether populations of some of the common butterflies are as sedentary as has sometimes been suggested, the sites in the monitoring scheme can reasonably be considered to have separate populations. Certainly, the rarer species in isolated biotopes occur as essentially independent population units. In discrete populations, the maintenance of synchrony over a large area, for many years, is surprising.

A common influence of weather, or other widespread factors, on generation-to-generation changes is not, by itself, likely to result in long-term synchrony. This can be understood if two isolated, but nearby sites of equal size are envisaged. These two hypothetical sites are identical except that one has an abundance of food plants in the conditions required for a particular butterfly, sufficient for many thousands of butterflies, and the other has only just sufficient appropriate food plants for 100 individuals of the species to survive. Assume that 10 mated females of the species are introduced to each site. The population at the first site is likely to increase rapidly, while that at the second site will, at best, just survive. Years of 'good' and 'bad' weather may be reflected to some extent in population changes from one year to the next at both sites, but the trajectories of the two populations will be quite different and, while this is true, synchrony will be virtually non-existent.

However if, after a period of several years, the increasing population at the ideal site reaches the carrying capacity of that site, and the individuals have to compete in some way for the larval food plants, the upward trend will end. Subsequently, if the second population has not become extinct, the fluctuations of both populations will be similarly affected by variation in weather conditions from year to year; a general synchrony of the fluctuations of the two populations will have been established, although their average numbers will remain very different.

Thus the synchrony of fluctuations in index values, observed in the data from the monitoring scheme, depends on the operation of a factor or factors that limit the growth of each population. We suggest that the most likely factor is a resource limit. No other factors such as the action of predators, disease, or parasitoids could, we believe, limit populations in so consistent a way at a large number of sites, often widely separated and in different types of countryside.

In particular, the frequent synchrony of populations suggests that fluctuations of these butterflies are not caused by interactions with local populations of predators, parasitoids, or competing herbivores. Thus the results from monitoring add to various other strands of evidence (e.g. Gilbert and Owen, 1990; Cornell and Lawton, 1992) to cast doubt on the importance of community interactions, historically one of the central tenets of population ecology.

The role of weather in the pattern of generation-to-generation fluctuations in monitoring data has been explored to some extent (Chapter 8). The effects are by no means restricted to direct impacts on the butterflies. There may be such direct effects, for example the restriction of egg-laying in cool or wet weather. There may also be a wide range of indirect effects, such as an indirect effect of weather on predation, which may be responsible for the associations between increased abundance and warm summer weather. Summer warmth may operate by speeding larval development and so reducing the time that the larvae are available for predation.

In summary, it is suggested that the synchrony observed in the monitoring data provides circumstantial evidence that many butterfly populations are limited, probably by the availability of larval resources, and only because of this limitation are the effects of weather broadly similar from site to site. Limiting factors are, by definition, density-dependent, acting more severely when the population is high relative to the abundance of resources. Direct tests for density dependence in population data are available, but because the butterfly data are index values such tests do not seem appropriate because of the possibility of non-linear (density-dependent) relationships between the indexes and population numbers.

14.2.2 Constancy of relative abundances and spatial distributions

Synchrony of population fluctuations must be reflected in stability in the relative size of index values at different sites; that is, only if populations at different sites increase or decrease by similar factors, is the difference between their index values, on which synchrony depends, maintained.

After many years of processing, analysing and interpreting data from butterfly transect counts, it is usually quite easy to identify sites in the monitoring scheme from a list of the species present and their index values in any year. In general, with some exceptions, the butterfly fauna of a site is a characteristic one; changes that do occur locally are often also widespread and affect all or most sites.

This stability is also present in the distribution of counts around transect routes within sites; examples given in Chapter 5 and in the site studies (Chapter 13) show that the distribution of counts is, for a given species, often very similar from year to year and the particular pattern may be maintained over a long period of years.

There is no doubt that these consistent spatial distributions are determined very largely by topography, shade, shelter and the structure, composition and flowering of vegetation. A clear example (Chapter 5) is the importance of shade to the distribution of butterflies in woodland. Each butterfly has its preferred shade conditions and these largely determine where in a wood it is seen in flight. As these physical features of the environment often remain more or less constant, or change slowly from year to year, the stability of local distributions is a natural consequence. In many cases, the spatial stability of adults will also indicate spatial stability of breeding areas; in butterflies which range more widely, the stability sometimes hides the fact that breeding areas have changed.

14.2.3 Departure from wider trends

Although, as has been described, long-term trends in local populations are often similar over large areas, in some cases gradual departures from wider trends occur. Again, examples are given in the site studies (Chapter 13).

These gradual changes in butterfly populations can usually be related to successional changes in vegetation. In conifer plantations, as the trees grow, there may be a decline of butterflies associated with open situations and an increase of those which prefer some shade. Similarly, in unmanaged grassland, butterflies such as the marbled white increase in numbers relative to wider trends as the grass grows tall and coarse, while other butterflies, such as the Adonis blue, decline.

The slow relative changes in index values are thus assumed to be associated with slow changes in physical factors, such as shade, or in resources (food plants) in the particular conditions required by a butterfly. The two factors, resource levels and the physical conditions in which they occur are, of course, intimately bound together and the butterfly may respond to broad physical conditions in order to find suitable larval food plants which occur in those conditions.

When the index values of a species at a particular site have been shown to increase or decrease, relative to wider trends, this has been considered an indication that a species is faring either better or worse than expected (Chapter 13). The inference made is that the vegetation of the site has changed and some factor required by the butterfly has been affected. In such assessments, a local departure of index values is treated as a point of interest, requiring explanation. However, left unmanaged, virtually all vegetation will show successional changes; thus long-term trends in butterfly abundance may be the norm, rather than the exception. The general synchrony of populations, emphasized in this discussion, may be related to the slow rate of change of vegetation or be a feature of managed countryside in which succession of vegetation has been suppressed. For example, regular grazing of grassland prevents it from developing into scrub and regular trimming of hedges prevents them from growing into lines of trees and, in each case, the suitability for particular butterflies is maintained.

When populations fall to a low level, after some catastrophic event, or when butterflies colonize, or are introduced into, a new area, there may be an opportunity to record the way in which populations behave when well below their resource limit. The number of such examples that have been monitored are few as yet, but there are clear indications of a rapid recovery of populations of several species, notably the speckled wood, ringlet and Adonis blue, following the severe declines after the 1976 drought. Similarly, the rapid increase, and subsequent stabilization, of the index values of the small skipper at Gibraltar Point (Chapter 7) has the classical features of growth towards an upper population limit.

In a relatively small number of instances, local populations of butterflies have changed rapidly and asynchronously. Such rapid changes are likely to be associated with radical management of sites. One example discussed in Chapter 12 showed the differing trends in three populations of the Adonis blue at a single site; these were interpreted as being caused by sharply differing grazing regimes in the areas occupied by the different populations.

There has also been little synchrony in the fluctuations of different populations of the heath fritillary in the coppice woods of Kent and its local distribution also changes rapidly (Chapter 12). These erratic, local increases and decreases in numbers are the result of very rapid changes in the shade and vegetation of small units of coppice woodland, following coppicing. Similarly, C.D. Thomas (1991) showed, using monitoring data, asynchronous changes in a population of the silver-studded blue in heathland on Anglesey. He concluded that the asynchrony was, as in the case of the heath fritillary, the result of rapid changes in the availability of suitable habitat. The heathers, on which the larvae of the silver-studded blue feed at this site, change, as they age, in their suitability for the butterfly. In this case, the cycle of growth is renewed by rotational cutting or burning. The transect counts of the butterfly for an individual section increased or decreased according to the age and condition of the heather in that section. In essence there were successional changes in vegetation, following management, analogous to those in coppice woodland.

Such rapid changes are not necessarily characteristic of a butterfly species *per se*, but of a species in a particular biotope. For example, populations of the heath fritillary in grassland change relatively slowly and synchronously, in contrast to their behaviour in coppiced woods. Similarly, in the newly created grassland of St Osyth (Chapter 13), butterflies such as the hedge brown, which generally fluctuate in broad synchrony over large areas of the country, increased rapidly in numbers, against the regional trends, as a rich flora rapidly became established.

Although the pattern of fluctuations in some of these examples differs sharply from what we have found to be the norm, the underlying reason seems to be the same; whereas the synchrony and spatial stability of most species at the monitored sites is a reflection of the stability of their resources, local variability of populations in a minority of cases is a reflection of a rapidly changing biotope. Thus these exceptions to the stability of spatial distributions reinforce the view that the character of the distributions depends on the physical character of the environment and, essentially, on the resources available to the butterfly.

The exceptions also suggest that the frequent synchrony of populations that has been found to occur may be, in large part, a feature of nature reserves which have been subject to relatively little drastic management and are changing only slowly. If the monitored sites included more farmland sites, where crops change, grazing animals move from field to field, hedges are cut intermittently and ditches drained, these factors might more often lead populations to fluctuate independently.

14.3 EVIDENCE FROM OTHER SOURCES

A wide and growing body of evidence, from studies of butterflies and other insects, suggests that resource availability plays a central role in their population ecology. The evidence can be divided into two main types, that from detailed studies of individual butterflies and more general considerations of the population ecology of insects.

Thomas (1991), in a major review of the ecology and conservation of rare butterflies, emphasized that the presence, even in great abundance, of the larval food plant at a site within the range of a species is no guarantee that it can breed there. Most butterflies have very specific requirements; females will usually lay their eggs only on food plants that occur in a restricted range of conditions. This applies to all of the rare butterflies discussed in Chapter 12 and also to many of the commoner species, although their requirements have been less studied. It is not always clear whether the selection by females of particular oviposition sites is related mainly to their ability to locate food plants quickly in such situations (as seems to be the case in the wood white, Chapter 12 and with the orange tip; Dempster, 1991) or whether larval survival is much higher on the plants selected. In any case, the implications are the same; that the availability of adequate resources cannot be equated simply with the availability of the food plant.

Although adult butterflies are also selective in their choice of flowers for feeding, most species will nevertheless feed from many different flowers as they become available (e.g. the meadow brown; Pollard, 1981). As far as we are aware, there is no evidence that shortage of flowers for adult feeding is important in limiting butterfly populations, except perhaps in extreme droughts.

In perhaps the classic example discussed by Thomas (1991), the very specialized requirements of the large blue are described. Although the association of the species with ants had been long known (Frohawk, 1924), the butterfly disappeared from many sites, including nature reserves, where both its food plant, thyme, and *Myrmica* ants remained abundant. Thomas (1980) showed that the larvae were even more specialized than had been thought; after feeding on thyme for a brief period, the larvae must be taken into the nests of only one species of *Myrmica* ant (*M. sabuleti*), as in nests of other species survival is too poor. Thomas (1991) writes that the carrying capacity, set by resources, is determined by '... the total number of *M. sabuleti* nests that have thyme within 2 m of their entrances'. The resources required by other butterflies may not be as precise as this, but in most cases they are much narrower than would be suspected from a superficial knowledge of their food plants and general biology.

Dempster (1983) reviewed detailed life-table studies of 24 species of butterflies and moths and found evidence for the action of 'density-dependent' factors, which might limit populations, in 16 of them. Given the difficulties of identifying limiting factors in short runs of data, this seems a sufficiently high proportion to confirm that such factors operate quite frequently. In 13 of the 16 cases, intraspecific competition for resources was detected; Dempster concluded that the evidence led him to favour a model of population limitation by a ceiling set by resources.

In one of Dempster's (1983) examples, that of the cinnabar moth, the resource level (the abundance of ragwort) fluctuated mainly as a result of variation in rainfall, and fluctuations in the abundance of the moth were largely caused by fluctuations in its resources. That is, in this case, resources not only limited the sizes of the populations but variation in resources determined the pattern of generation-to-generation fluctuations.

In the case of the orange tip butterfly in Monks Wood (Dempster, 1991), fluctuations in abundance of eggs were found to be closely correlated with abundance of suitable food plants. In this wood the only food plant is lady's smock, and the eggs are laid on flowering stems in the spring. Like the rare butterflies, the orange tip was very selective and only large plants in sunny and sheltered situations were used. Females tended to avoid plants on which an egg had recently been laid by another butterfly, a form of indirect competition, and if they failed to find suitable food plants they appeared to leave the wood. By this sort of intensive study of an insect and its resources, it becomes possible not only to demonstrate limitation by resources but also to show some of the mechanisms by which it operates.

Dempster and Pollard (1981) gave several further examples in which population fluctuations of insects followed closely the fluctuations in their resources. It is possible that some of the butterfly populations we have monitored behave similarly, for example the heath fritillary, discussed in this chapter, and other fritillaries associated with the rapidly changing environment of coppice woodland. We do not, however, suggest that all, or even most, butterfly populations are responding to resource fluctuations in such a direct and immediate way. In many species it is likely that the availability of resources sets a limit to population size and the butterflies fluctuate in numbers below this limit; even in such cases, however, occasional events such as extreme drought may lower the resource limit to such an extent that the butterfly population is directly affected.

Interpretation of the patterns of fluctuations in numbers of moths, caught at light traps throughout Britain, has been very different from that presented here (e.g. Taylor, 1986). Taylor has emphasized the fluidity of spatial distributions of moths, considered at a national scale, and the importance of the movement of individuals in causing this fluidity. In contrast, we have emphasized stability. Whether butterflies and moths differ radically in their population dynamics, or whether their dynamics are essentially similar, but are being interpreted differently, is unclear.

14.4 A FRAMEWORK FOR BUTTERFLY POPULATIONS

In this section, in summarizing the contents of this chapter, we place the ideas developed against the more general background of population theory and suggest a framework for butterfly populations.

14.4.1 Background

Population ecology has long been, and remains, a controversial subject. It would not be appropriate here to discuss the competing theories on the ways in which populations may interact with each other and with their environments. Instead our own views, based on the evidence discussed in this chapter and on other more theoretical considerations, are given as briefly as possible; we recognize that by no means all population ecologists will agree with them.

It is conventional to divide the many factors which affect the size of populations into those which are independent of population density (density-independent) and those which act more severely as population size increases (density-dependent). The density-independent factors are usually considered to cause populations to fluctuate from generation to generation, while the density-dependent factors are thought of as dampening, regulating or limiting (these words are often used as alternatives) the fluctuations.

In practice the distinction between density-independence and density-dependence may not always be sharp; many factors, such as disease or predation may be density-dependent over a restricted range of population sizes or for a limited number of generations and be density-independent at other times. In addition, a density-dependent relationship may be present, but weak and variable and scarcely distinguishable in its impact on a population from a strictly density-independent factor. We accept Milne's (1957a, b, 1962) view that because of the inherent variability of nature, only competition for resources acts in a sufficiently consistent (inevitable) density-dependent way to provide an effective limit to the size of populations. The word 'limiting' seems to us more appropriate to this view of population processes, than 'dampening' or 'regulating', because it is only the growth of populations that is halted; there is no equivalent buffer to keep declining populations from extinction.

This view of population processes is essentially that proposed originally by Milne and supported by Dempster (1975). This view does not deny the importance of predators, parasitoids, diseases and other factors as major causes of mortality, but suggests that they very rarely provide a consistent limit to population size.

14.4.2 A framework for butterfly populations

For butterflies, the view of population dynamics outlined above translates to the following.

1. The presence of a butterfly species at a particular site depends on a wide range of factors. Some that are likely to be involved are the abundance of food plants in conditions suitable for egg-laying, the abundance and quality of flowers for adult feeding, the severity of weather conditions, the abundance of predators and parasitoids and the prevalence of diseases. In order to persist in an area, a population must have the capacity to grow in the presence of these factors and any others that have an impact on the population.
2. The availability of resources is, for most butterfly species, the major factor which limits population growth and determines mean population size. The limiting resources are usually larval food plants in the particular conditions required by a species. These requirements may be very precise and cannot be equated directly with the abundance of the larval food plants.
3. Generation-to-generation fluctuations in population size may be caused by any of the range of factors listed in (1). Two of these factors are likely to be of special importance in this respect. First, variation in weather is likely to have a major effect because it can act directly on individuals in a population or indirectly through its effect on any one of the other factors. Secondly, the availability of resources not only sets an upper limit to population size, but variations in that limit may sometimes override other factors and so determine generation-to-generation population fluctuations.
4. Many butterflies, especially the common and widespread species, occur in more or less interconnected populations. It is likely that, in any one year, only some of these populations reach their resource limit; these 'successful' populations may produce a surplus of individuals which disperse to, and so sustain, the less successful populations.

– 15

Climatic warming

15.1 INTRODUCTION

There seems little doubt that, over the next few decades, the increase of carbon dioxide and other 'greenhouse' gases in the atmosphere will lead to global warming. Carbon dioxide concentration has already increased from a pre-industrial level of about 270 ppmv (parts per million by volume) to the present 350 ppmv and may reach 540 ppmv by the middle of the next century (DoE, 1988).

Estimates of the likely increase in global temperature vary considerably, but most are of the order of 1–4°C by the year 2050. The decade of the 1980s was the warmest of the century, at the global level, and it is possible, although by no means certain, that climatic warming has already begun. Although global temperatures are likely to rise, changes in particular regions of the world are more uncertain and some areas may even become cooler. A recent estimate for Britain (DoE, 1988) indicated a mean temperature rise of around 3°C and a change in mean rainfall of ± 20%. All that is assumed in this chapter is that the British climate will become warmer and that rainfall patterns may change.

There have been previous substantial temperature changes in the Post Glacial period. Pollen analysis and the study of animals such as beetles and snails, which leave abundant remains, have shown that the distribution of both plants and animals has changed in response to these climatic changes. There is no equivalent fossil record of butterflies, but Dennis (1977) has reconstructed the likely course of their colonization of Britain after the last ice age and changes in distribution in subsequent periods of climatic change. The predicted increase in temperature in the next century is as great as or greater than in these earlier periods, and the predicted rate of change more rapid than has occurred in the Post Glacial. Thus some of the flora and fauna may be less able to respond by movement or adaptation than in earlier warm periods.

Although many factors may be involved in determining the ranges of butterflies, the strong connection between butterflies and climate has been recognized by many authors. In addition to the historical evidence (Dennis,

1977), Heath *et al.* (1984) began their summary chapter of the *Atlas of Butterflies in Britain and Ireland*, which describes recent distributions, by writing 'Climate and geology set limits to the maximum range which can be attained by our butterflies.' More recently, Turner (1986) and Turner *et al.* (1987) have drawn attention to the strength of the relationship between energy input from the sun and the richness of the butterfly fauna.

The effects of climatic warming are likely to be very damaging to agriculture and wildlife. Traditional patterns of agriculture will undoubtedly be disrupted; for example, the main cereal growing areas of the world are expected to shift northwards, with drought becoming an increasing problem in the present cereal belts. Clearly, such changes will not be achieved smoothly and an increase in the frequency of famines in various parts of the world seems inevitable. A further serious problem is the expected rise in sea level, with the almost inevitable inundation of some heavily populated parts of the world and serious problems for most coastal industry and habitation.

Set against these potentially catastrophic changes, concern about the effects of climatic warming on butterflies may seem absurd. However, in addition to direct effects on humans and their dwellings, industry and agriculture, climatic warming will have a major impact on the natural and semi-natural biotopes of the world, of which butterflies are a part. These changes too will have their implications for the economic activities of humans. Even where economic interests are not directly threatened, as the dominant animal, humankind now has a clear responsibility to safeguard the diversity of life on earth.

As described in the Preface to this book, the fauna of Britain has relatively few species, even as compared with the rest of Europe. The distribution of many groups in Britain is well known, largely through work co-ordinated by the Biological Records Centre of the Institute of Terrestrial Ecology. Trends and fluctuations in numbers of birds (e.g. Marchant *et al.*, 1990), moths (e.g. Woiwod and Dancy, 1987) and butterflies are much better known than elsewhere in the world. Britain, therefore, is uniquely placed to monitor the effects of climatic warming on its animal life.

A broad review of climatic change and the British butterfly fauna is provided by Dennis and Shreeve (1991). In this chapter, the emphasis is on the role of monitoring in relation to climatic warming. This role has two main elements: first, the existing data may be used to develop predictions of likely short-term effects; second, continued monitoring may detect changes in abundance and distribution at an early stage, so that any possible conservation measures may be considered.

15.2 USE OF EXISTING MONITORING DATA

The associations between weather and abundance, shown by the monitoring results, provide the best available information for prediction of effects of climatic warming on butterflies, and their value will increase over the years. However, it is unlikely that the results obtained from monitoring over just 15 years can predict these effects with any precision. First, the short period of recording means that relationships with weather can be shown only imprecisely and in a simplified way. Second, the likely increase in temperature, and perhaps changes in rainfall and other aspects of weather, will take these climatic factors far outside the ranges that have occurred in these 15 years. The responses of butterflies in these new conditions may be similar in general character to those that have been recorded, but the detailed nature of the responses may be very different.

The associations between butterfly abundance and weather (Chapter 8) suggest that some species have responded to warm summers by increasing in abundance, with no evident harmful effects of the dry conditions which often accompany warm weather. In many species, such as the small skipper, marbled white and hedge brown, the increases occurred during each warm summer, but in other cases, such as the grizzled skipper and brimstone, the increase was in the following year.

A few species may respond dramatically to just two or three warm summers; the recent increase in abundance and expansion of range of the holly blue is thought to give an indication of the speed of response that is possible (Chapter 11). It seems likely that increased abundance will often be associated with changes in range, but one cannot assume that this will always be true. For instance, in the case of the hedge brown (Pollard, 1991b, 1992), described in Chapter 10, the recent expansion of range has not been associated with any clear increase in abundance.

In contrast, some species, while benefiting from warm weather, seem prone to harmful effects of drought. Notable amongst these species are the green-veined white and speckled wood, butterflies which usually live in relatively cool and shady situations, but the data for other species, such as the common blue and Adonis blue, suggest similar effects. The sites at present occupied by the Adonis blue, which is at the northern edge of its range in Britain, are mainly on south-facing slopes with shallow soils; on such sites the food plants are particularly vulnerable to drought. Similar considerations apply to several butterfly species on grassland and heathland nature reserves in southern Britain.

The main effects of drought are probably caused largely by desiccation of food plants and consequent starvation of larvae, although the adults may suffer if they can not obtain sufficient food and moisture from flowers or elsewhere. Often, drought is the result of a combination of lack of rainfall and high temperature, the latter increasing evaporation and transpiration. Thus, even if rainfall in Britain does not become lower as the climate becomes warmer, effects of drought could become more serious.

Although we suggest a strong impact of drought, the effects are likely to vary over the country. While droughts may make the south and east of Britain less favourable for butterflies, there may be few such problems in the north and west, and in these areas there may be an uncomplicated benefit to many butterflies from warmer summers. This supposition can be tested by further analyses of monitoring data, in particular of geographical differences in responses to weather at individual monitored sites.

15.3 MONITORING AND LONG-TERM EFFECTS OF CLIMATIC WARMING

The particular conditions in which their food plants occur are of great importance to butterflies (Thomas, 1984, 1991). One effect of a rise in temperature may be to make a larger proportion of the food plants available to the butterfly. A simple example, discussed in Chapter 12, is that of the Adonis blue. This species may no longer be mainly restricted to south-facing slopes and may be able to flourish where horseshoe vetch occurs on cooler slopes and in longer vegetation. Indeed, the slopes where the species now occurs may become unsuitable. If such changes occur, they may result in dramatic changes in the abundance of local populations; species may disappear from some sites and colonize new ones. The balance of losses and gains may depend on the ability of the butterfly to reach new sites. Distributional changes of these types should be readily detected by monitoring. Within a site, the distribution of counts around the transect route may change. If extinctions occur and new sites are colonized, these changes will also be detectable by methods such as those discussed in Chapter 7.

Thus, in the short term, increased temperatures may change the ability of butterflies to exploit existing biotopes. In the longer term the biotopes themselves will change in character and this will also have effects on butterfly distributions. Climatic warming will inevitably result in gradual changes in the ranges of many plant species. Changes in local distributions, and also expansions and contractions of range of the food plants of butterflies will occur. Biotopes, such as woodland, may gradually change in character and provide conditions that exclude some of the present butterflies and permit others to colonize them. Interpretation of such changes in distributions of butterflies will depend on effective monitoring of vegetation. At present, there is little or no monitoring of vegetation at sites in the Butterfly Monitoring Scheme and this is an area which needs to be improved.

Dennis and Shreeve (1991) suggest that the northern butterflies, such as the large heath and mountain ringlet, may be particularly vulnerable to climatic warming. It seems likely that their populations will become more isolated as the areas of upland bogs and montane grassland become more restricted. Monitoring of such areas needs to be increased in spite of the difficulties posed by inaccessible terrain and the scarcity of recorders.

The butterflies will themselves change as they adapt to the new conditions. There seems little point in speculating on the nature of these adaptive changes in any detail; the possibilities are manifold. However, some clues may be obtained from studies of those British butterflies which now occur in substantially warmer and drier regions of the world. The first adaptations to be detected will be those that are measured easily and have already been studied extensively. Examples include wing morphology (e.g. Brakefield, 1984; Dennis and Shreeve, 1989), the voltinism of those species with a flexible number of generations (e.g. Lees, 1969; Nylin, 1989) and perhaps the length of the flight-period (Brakefield, 1987; Pollard, 1991b). The last two of these may be detected by monitoring.

The work of Lees (1962, 1969), Lees and Archer (1980) and Pullin (1986) has suggested that the voltinism of several butterfly species can change rapidly in response to artificial selection and, presumably, similar changes would occur in response to climatic change. In species such as the small heath and common blue, voltinism is flexible; in warm years more individuals produce an additional generation. In addition, it seems likely that a series of warm years will, by selection, result in an increased disposition of populations to produce additional generations. For example, the common blue in Scotland, where it is normally univoltine, produces a small second generation in exceptionally warm summers (Chapter 11). If individuals of this second generation breed successfully in a succession of warm summers, selection may eventually lead to the production of a regular second generation, as is the case farther south.

Changes in voltinism with rising temperature are likely in many species and their current voltinism, farther south in Europe, provides some guide to this. The peacock, for example, is bivoltine in southern Europe and may well become so in Britain. However, many species are univoltine throughout their range and the adaptations to a warmer climate may take different forms. The meadow brown female, in Mediterranean regions, emerges in the spring, mates and then has a long summer aestivation, becoming active again in the autumn when the eggs are laid (Scali, 1971). In this way the hot, dry weather of summer is avoided and the larvae are more likely to find palatable grasses in the cooler and moister autumn weather. On chalk grasslands in Britain, the flight-period of the meadow brown (Chapter 10) is so long that the first individuals emerge in June, but the last eggs are laid as late as October. Summer drought on these shallow chalk soils may have had a role in the evolution of this long flight-period (but see Shreeve, 1989;

Chapter 10). Similarly, in Spain the grayling female lives for much longer than in Britain (Garcia-Barros, 1988); maturation of the eggs is delayed, and so a dry mid-summer period may be avoided.

One indication of a change in the flight-period of a butterfly has already been detected in monitoring scheme data (Pollard, 1991b). The flight-period of the species concerned, the hedge brown, has become longer and perhaps slightly earlier during a period of expansion of range (Chapter 10). Although climatic change did not appear to be responsible for the changes, the example does indicate the flexibility of the flight-period and the potential of the monitoring scheme for detecting such changes.

Climatic change is not the only likely result of raised levels of atmospheric carbon dioxide. The growth of many native plant species, under experimental conditions, is increased by the enhancement of photosynthesis (e.g. Hunt *et al.*, 1991). Experimental studies have shown that such increased plant growth may not result in more rapid growth of caterpillars (e.g. Fajer *et al.*, 1989) because the carbon/nitrogen ratio of the plant material increases; thus more must be consumed to ingest a given amount of the nitrogen required for proteins. However, amounts of nitrogen deposited from the atmosphere have also increased and, in combination with increased carbon dioxide, could benefit some herbivorous insects. Taking speculation even further, it is not impossible that an extended growing season of some grasses could lead to an extension of the period in which they are suitable for feeding by caterpillars and so a lengthening of the adult flight-period, as found in the hedge brown (Chapter 10).

Bobbink (1991) has suggested that the increased dominance of tor grass in Holland may have been caused by input of atmospheric nitrogen. This grass has also become more abundant on downland in England, resulting in the suppression of characteristic downland herbs, with adverse effects on some butterflies. Bobbink's evidence of an effect of nitrogen on tor grass was based on experiments with fertilizers; thus the role of atmospheric nitrogen has not been established. However, the study serves as a reminder that, while climatic warming lies in the future such direct impacts of atmospheric carbon dioxide and nitrogen on plants could have been in operation for many years.

Climatic warming will undoubtedly mean that Britain will become suitable for some continental butterflies. The map butterfly, *Araschnia levana*, survived here, following introduction, for at least 2 years early in this century (discussion by Emmet and Heath, 1989). This species has spread both southward and northward in Europe in recent decades and may be a contender for immigration and establishment in Britain, even without climatic warming. Another butterfly, the continental subspecies of the swallowtail, *Papilio machaon gorganus*, has been established in Kent for short periods on several occasions in the last 150 years; there is little doubt that this will be one of the first immigrants to establish in Britain if temperatures rise.

It is very unlikely that the first arrivals in Britain would be detected at monitored sites, but it is likely that some of the first populations will establish on the monitored nature reserves in southern England. Comprehensive study of the establishment and build-up of populations will depend, in part, on systematic monitoring.

– 16

Synopsis

16.1 INTRODUCTION

In this final chapter, we look back over the past 16 years of monitoring and draw some conclusions from the results and from the experience of monitoring; then we look forward to consider some of the problems facing the butterflies in Britain and the role which monitoring might have in their conservation.

16.2 SUMMARY OF THE MONITORING PERIOD

Since its inception in 1976, the Butterfly Monitoring Scheme has provided information on the changes in numbers of butterflies at well over 100 sites in Britain. The information has been used by conservation agencies to monitor populations of rare and endangered butterflies and, in some cases, to assess the effectiveness of management implemented for their conservation. Many sites are monitored independently by local conservation bodies and, in many of these cases, comparative use has been made of information from the national scheme. In addition, individual research projects, such as that at the landfill site at St Osyth, described in Chapter 13, have made use of data from the scheme for comparison with local results.

The wide geographic range of sites and the variety of their biotopes have helped to provide a unique body of data on the commoner butterflies. This information ranges from the patterns and extent of population fluctuations, to the length and timing of flight-periods, local distributions, extinction of local populations, the foundation of new ones, and migratory movements. For most of these topics no equivalent data are available for butterflies in Britain or elsewhere in the world.

The pattern of population fluctuations, together with other evidence from population studies of butterflies, has been interpreted to suggest that butterfly populations are generally limited in size by the availability of suitable food resources for the larvae. If this interpretation is correct, it provides an ecological basis for the conservation of butterflies through the enhancement of their resources by management.

Below the resource limit, generation-to-generation fluctuations in population size appear to be strongly influenced by weather. A few, mainly bivoltine butterflies, seem to undergo frequent local extinction and colonization which are related to general abundance and so to weather. These findings have clear importance for the prediction of effects of climatic warming.

The long-term trends show that some common butterflies have increased in abundance at sites in the scheme. As several of these species have expanded their ranges in the same period and some have colonized new sites in the monitoring scheme, there is little doubt that these butterflies are at present flourishing. It is possible that these changes are related to weather, but no clear link has yet been demonstrated. In contrast to these increases, there is some evidence from the monitoring scheme, albeit limited, that populations of rare butterflies have continued to decline. Evidence from other sources suggests that continued declines of some of the rare butterflies have been severe.

Flight-periods have been shown to vary considerably from year to year (Chapter 10), depending on summer temperature. In warm seasons, when the flight-period is early, population numbers often increase, so that phenology and population dynamics are linked. Although flight-periods are earlier in warm seasons than in cold seasons, there is, for many species, little indication in any given year that flight-periods are earlier in the warmer south of the country than in the north. This result suggests that for these species the timing of the flight-period is very important and that the butterflies may be able to adjust their growth rates so that they emerge as adults at an optimum date. The study of the phenology of butterflies, through the monitoring scheme data on flight-periods and by other means, may help in determining how the ranges of butterflies are limited.

The information from the monitoring scheme, together with the essential information on population processes from detailed autecological studies, promise to provide a broad picture of the population ecology and biogeography of butterflies which is unrivalled for any other group of insects.

16.3 THE FUTURE OF BUTTERFLY CONSERVATION AND THE ROLE OF MONITORING

At the start of this book, attention was drawn to a dichotomy in the distribution of British butterflies; this dichotomy is particularly evident in lowland, cultivated Britain. In general, in such areas, the butterflies divide quite sharply into the rare localized species and those that are common and widespread. There is not, as might be expected, a continuum of increasing frequency of occurrence from the rarest species to the commonest; rather, butterfly species are either common and widespread or rare and localized.

The rare butterfly species are more or less confined to restricted areas which provide the special resources that they require. They may be abundant at such sites, but are rarely seen away from them in the farmed landscape. These are the butterflies which, to a large extent, depend on nature reserves and other protected areas for their survival. In contrast, the common butterflies may be found in uncultivated corners of farmland, along road verges and green lanes, often in gardens, in many grasslands and in woodland rides, and in other fragments of uncultivated land. The dichotomy presents problems in monitoring British butterflies (Chapter 4) and is also a danger signal for the future of the rare species.

Thomas (1991) has drawn attention to an apparent paradox in the ecology of the rarer butterflies. Many of these species, such as the silver-spotted skipper, small blue, Adonis blue, chalkhill blue and the woodland fritillaries, are associated with early stages in the succession of vegetation. In most of these cases, their habitats can be maintained by management, and the sedentary nature of the butterflies is not then such a disadvantage. However, as Thomas (1991) points out, their habitats in the natural vegetation of Britain would have been clearings in a largely wooded countryside, and mobility is essential to species which exploit such ephemeral conditions. He suggests that all of these rare butterflies were once much more mobile than they are now, and that there has been selection against mobility as their populations have become more isolated. If the nearest suitable sites are tens of kilometres away, mobile individuals which fly away from such isolated populations are unlikely to survive to leave progeny.

Once a butterfly species has declined to such an extent that it occurs only in isolated populations of sedentary individuals, active conservation may then be necessary to keep it from extinction. Arnold (1980) came to an identical conclusion about rare butterflies in California; there is little doubt that the same is true in other areas of the world where natural or semi-natural biotopes have become isolated by intensive land use. In southern Britain, several rare butterflies have reached this vulnerable stage and it is probable that their survival will depend on the continuation of appropriate management of nature reserves. In parts of the north of Britain, populations of such species as the large heath and northern brown argus may be approaching a similar state.

For such isolated populations in particular, climatic warming is a danger. It is possible that some of the species would benefit because higher temperatures might enable them to exploit a wider range of sites, in the ways described in Chapter 15. It is equally possible that, because of their lack of mobility, they would fail to find these newly suitable areas before becoming extinct at their existing sites.

In theory, the provision of a network of sites with semi-natural vegetation, throughout the intensively cultivated areas of the countryside, might aid the dispersal of butterflies. However, at present little is known of the ways in which butterflies disperse and thus how their dispersal might be aided most effectively. It is likely that conservation of some of the rare species will require the transfer of individuals to new sites.

For the rare butterflies, the need for monitoring is clear, and the existing Butterfly Monitoring Scheme seems appropriate. It is based to a large extent on the nature reserves where they occur. In addition, many other reserves are monitored by similar methods. Whether or not climatic warming adds a further dimension to the problems of conserving these butterfly species, monitoring of many populations of rare butterflies should continue.

The current expansion of range of several of the common butterflies is intriguing. These expansions may be directly related to climate but this has yet to be demonstrated clearly. In Chapter 8, population models which incorporate weather variables were described. As data for more years become available, these models should become more precise in their description of the effects of weather on butterfly populations. They could then also be used with historic weather data to reconstruct the likely course of past changes in butterfly abundance. Thus, the models could test whether the contractions of range of many common butterflies in the latter part of the nineteenth century and expansion in the latter half of this century might reasonably be explained by weather.

The apparent increase in the flight-period of one of the common butterflies, the hedge brown, at a time of expansion of range, is also intriguing. This too is an area that requires further study, both of the flight-period of the hedge brown and those of other butterflies; the data are already available for such studies to begin.

16.4 INTEGRATION OF STUDIES ON BUTTERFLIES

There is no doubt that if the Butterfly Monitoring Scheme had been conducted in isolation from other studies of butterflies, its value would have been greatly diminished. It has benefited greatly from the detailed distribution data that are available for British butterflies and in particular from the, relatively few, detailed population studies that have been conducted on the rarer butterflies.

Both monitoring abundance and mapping distributions are considered to provide a good return from scientific funding, because they provide information on many species. In contrast, a detailed study of a butterfly population may require 5–10 years' work and will provide information only on one species at one site. Nevertheless, such studies are vital for an understanding of the population ecology and requirements of species, and to enable full use to be made of monitoring and mapping. There is a clear need for the continuation of these different types of studies, for their closer integration, and for the identification of major gaps in current knowledge.

16.5 FINAL THOUGHTS

One unexpected bonus of monitoring has been an increased interest in butterflies and their requirements, both by the individuals who monitor them (including many nature reserve wardens) and by conservation organizations. Thus, in addition to the information that monitoring has provided, the very process of monitoring may have contributed to butterfly conservation.

Appendix A
Life cycles, food plants and behaviour

Interpretation of the data obtained from monitoring butterflies depends on a knowledge of their natural history and biology. Here we present, mostly in the form of a table (Table A1), some of the information most relevant to butterfly monitoring. For further reading, several books are listed at the end of this appendix. Other studies quoted here are included in the main list of references.

British butterflies vary considerably in the details of their life cycles. For example, overwintering of different species may be as an egg, larva, pupa or adult. The number of generations a year may be fixed at one or two, or be more flexible and vary according to the warmth of the season and the geographical location, as discussed in Chapter 10.

The food plants of the larvae of British butterflies are now reasonably well known, although some gaps in information remain. In particular, the range of species which some grass-feeding butterflies use is not fully known.

The behaviour of adult butterflies affects their chance of being seen and counted. A few butterflies, such as some of the hairstreaks, fly mainly in the canopy of trees and are seen only occasionally at ground level. Even those butterflies which fly amongst low vegetation vary in the frequency with which they are seen, because of their colour, size, speed and manner of flight and other aspects of their behaviour.

Male butterflies in particular vary in their behaviour from species to species. Some males, such as those of the peacock, are very conspicuous; they 'perch' at vantage points from which they fly up to investigate passing butterflies and intercept females of their own species. The vantage points they select are not permanent territories, but are occupied for a few hours or perhaps days, before the butterfly moves elsewhere (Baker, 1972a). Such behaviour occurs not only in butterflies, such as the brown argus and green hairstreak, which spend their lives within a small area; even the painted lady, a trans-continental migrant, adopts this strategy for mate location (Thomas, 1986).

Males of other species, such as the orange tip and wood white, actively search for females by almost constant flight, behaviour termed 'patrolling'. In

Table A1 Summary of phenology, voltinism, mobility, courtship behaviour and main food plants of butterflies in Britain. Information based mainly on Heath *et al.* (1984), Thomas (1984), Emmet and Heath (1989), Thomas and Lewington (1991) and Dennis (1992). Months: March (Ma), April (Ap), May (M), June (J), July

		Overwinter stage
Hesperidae		
Carterocephalus palaemon	Chequered skipper	Larva
Thymelicus sylvestris	Small skipper	Larva
Thymelicus lineola	Essex skipper	Egg
Thymelicus actaeon	Lulworth skipper	Larva
Hesperia comma	Silver-spotted skipper	Egg
Ochlodes venata	Large skipper	Larva
Erynnis tages	Dingy skipper	Larva
Pyrgus malvae	Grizzled skipper	Pupa
Papilionidae		
Papilio machaon	Swallowtail	Pupa
Pieridae		
Leptidea sinapis	Wood white	Pupa
Colias croceus	Clouded yellow	–
Gonepteryx rhamni	Brimstone	Adult
Pieris brassicae	Large white	Pupa
Pieris rapae	Small white	Pupa
Pieris napi	Green-veined white	Pupa
Anthocaris cardamines	Orange tip	Pupa
Lycaenidae		
Callophrys rubi	Green hairstreak	Pupa
Thecla betulae	Brown hairstreak	Egg
Quercusia quercus	Purple hairstreak	Egg
Strymonidia w-album	White-letter hairstreak	Egg
Strymonidia pruni	Black hairstreak	Egg
Lycaena phlaeas	Small copper	Larva

(Jy), August (A), September (S), October (O). The behavioural terms are discussed in the text: pe, perch; pa, patrol; agg, aggregations. In a few cases, marked ?, no clear information has been found

Generations	*Flight-period*	*Mobility*	*Male behaviour*	*Food plants*
1	MJ	Sedentary	Perch	*Brachypodium sylvaticum* *Molinia caerulea*
1	J–A	Sedentary	Patrol	*Holcus lanatus*
1	JyA	Sedentary	Patrol	*Dactylis glomeratus* *Holcus mollis*
1	Jy–S	Sedentary	?	*Brachypodium pinnatum*
1	AS	Intermediate	Perch	*Festuca ovina*
1	JJy	Sedentary	pe/pa	*Dactylis glomerata* *Molinia caerulea*
1	MJ	Sedentary	Perch	*Lotus corniculatus* *Hippocrepis comosa*
1	MJ	Sedentary	Perch	*Fragaria vesca* *Potentilla* spp.
1	MJ	Intermediate	Patrol	*Peucedanum palustre*
1	MJ	Intermediate	Patrol	*Lathyrus pratensis* and several other legumes
C	Ma–O	Wide-ranging	Patrol	Clovers, lucerne
1	Ap–J/Jy–S	Wide-ranging	Patrol	Buckthorns
2	MJ/JyA	Wide-ranging	Patrol	Cultivated brassicas
2	MJ/JyA	Wide-ranging	Patrol	Cultivated brassicas
2	MJ/JyA	Intermediate	Patrol	*Cardamine pratensis* and other crucifers
1	Ap–J	Intermediate	Patrol	*Cardamine pratensis* and other crucifers (flower-heads)
1	MJ	Sedentary	Perch	*Helianthemum chamaecistus* various legumes and flowers of shrubs
1	AS	Intermediate	agg	*Prunus spinosa*
1	JyA	Sedentary	?	*Quercus* spp.
1	JyA	Sedentary	?	*Ulmus* spp.
1	JJy	Sedentary	?	*Prunus spinosa*
2	MJ/JyA	Intermediate	Perch	*Rumex* (sorrels)

Table A1 Continued

		Overwinter stage
Lycaena dispar	Large copper	Larva
Cupido minimus	Small blue	Larva
Plebejus argus	Silver-studded blue	Egg
Aricia artaxerxes	Northern brown argus	Larva
Aricia agestis	Brown argus	Larva
Polyommatus icarus	Common blue	Larva
Lysandra coridon	Chalkhill blue	Egg
Lysandra bellargus	Adonis blue	Larva
Celastrina argiolus	Holly blue	Pupa
Maculinea arion	Large blue	Larva
Riodinidae		
Hamearis lucina	Duke of Burgundy	Pupa
Nymphalidae		
Ladoga camilla	White admiral	Larva
Apatura iris	Purple emperor	Larva
Vanessa atalanta	Red admiral	Adult
Cynthia cardui	Painted lady	–
Aglais urticae	Small tortoiseshell	Adult
Nymphalis polychloros	Large tortoiseshell	Adult
Inachis io	Peacock	Adult
Polygonia c-album	Comma	Adult
Boloria selene	Small pearl-bordered fritillary	Larva
Boloria euphrosyne	Pearl-bordered fritillary	Larva
Argynnis adippe	High brown fritillary	Egg
Argynnis aglaja	Dark green fritillary	Larva
Argynnis paphia	Silver-washed fritillary	Larva
Euphydryas aurinia	Marsh fritillary	Larva
Melitaea cinxia	Glanville fritillary	Larva
Mellicta athalia	Heath fritillary	Larva
Satyridae		
Pararge aegeria	Speckled wood	Larva/pupa
Lasiommata megera	Wall	Larva
Erebia aethiops	Scotch argus	Larva

Generations	*Flight-period*	*Mobility*	*Male behaviour*	*Food plants*
1	JyA	Sedentary	Perch	*Rumex hydrolapathum*
1	MJ	Sedentary	agg/pe	*Anthyllis vulneraria*
1	JyA	Sedentary	Patrol	*Ulex europaeus, Erica* spp.
1	JA	Sedentary	?	*Helianthemum chamaecistus*
2	MJ/JyA	Intermediate	pe/pa	*Helianthemum chamaecistus*, some cranesbills
2 1 (north)	MJ/JyA J–A	Intermediate	pe/pa	*Lotus corniculatus* and some other legumes
1	Jy–S	Sedentary	Patrol	*Hippocrepis comosa*
2	MJ/AS	Sedentary	Patrol	*Hippocrepis comosa*
2	AM/JyA	Wide-ranging	?	*Ilex aquifolium, Hedera helix* and other shrubs
1	JJy	Sedentary	Patrol	*Thymus drucei* and ant larvae
1	MJ	Sedentary	Perch	*Primula veris* and *P. vulgaris*
1	J–A	Intermediate	Perch	*Lonicera periclymenum*
1	JyA	Intermediate	agg.	*Salix caprea*
–	–	Wide-ranging	Patrol	*Urtica* spp.
–	–	Wide-ranging	Perch	*Cirsium* spp., *Carduus* spp.and some other plants
2	ApM/JJy/SO	Wide-ranging	Perch	*Urtica* spp.
1	MaAp/J–S	Wide-ranging	Patrol	*Ulmus* spp.
1	MJ/JyA	Wide-ranging	Perch	*Urtica* spp.
2	ApM/JJy/S/O	Intermediate	Perch	*Urtica* spp., *Ulmus* spp.
1	MJy	Sedentary	Patrol	*Viola* spp.
1	MJ	Sedentary	Patrol	*Viola* spp.
1	JyA	Intermediate	Patrol	*Viola* spp.
1	J–A	Intermediate	Patrol	*Viola* spp.
1	J–A	Intermediate	Patrol	*Viola* spp.
1	M–Jy	Intermediate	Patrol	*Succisa pratensis*
1	MJ	Intermediate	Patrol	*Plantago lanceolata*
1	JJy	Sedentary	Patrol	*Melampyrum pratense* and some other herbs
2*	Ap–O	Sedentary	pe/pa	Many grasses
2	MJ/A	?	pe/pa	Many grasses
1	Jy–S	Sedentary	pe/pa	*Molinea caerulea* and probably other grasses

Table A1 Continued

		Overwinter stage
Erebia epiphron	Mountain ringlet	Larva
Melanargia galathea	Marbled white	Larva
Hipparchia semele	Grayling	Larva
Pyronia tithonus	Hedge brown	Larva
Maniola jurtina	Meadow brown	Larva
Aphantopus hyperantus	Ringlet	Larva
Coenonympha pamphilus	Small heath	Larva
Coenonympha tullia	Large heath	Larva

these species, patrolling appears to be the only method of finding females, but other species, including the speckled wood (Shreeve, 1984) and wall (Dennis, 1982–83), change their behaviour according to the temperature. When the temperature is low, individuals are more likely to perch at a vantage point as they then benefit most from the warmth of the sun; as the temperature rises, flight is easier and patrolling more common. The density of males and availability of females also seems to influence the type of behaviour.

A third pattern of mating behaviour is that shown by some of the butterflies of the tree canopy, such as, in Britain, the brown hairstreak and purple emperor. In these butterflies, the males and females of a colony aggregate for mating at a particular tall tree. This behaviour improves the chance of finding a mate in species which are normally widely dispersed at low density.

In general, male butterflies are more active than females and are more likely to be seen. Males of most species spend more time in flight, while females may feed more at flowers and be otherwise inconspicuous, except when searching for places to lay eggs. In addition, males of several butterflies, especially the blues (Lycaenidae) are more conspicuous than the females. Thus, while uneven sex ratios certainly occur amongst butterflies, some which are apparently male biased may not in fact depart from a 1:1 ratio.

The flight behaviour of individual butterflies varies, not only in the details of mate location and the search for oviposition sites, but in the distance they cover during a lifetime. At one extreme, a black hairstreak may spend its life around a few blackthorn bushes in a woodland clearing (Thomas, in Emmet and Heath, 1989); at the other extreme, the painted lady seems to be capable of flight from North Africa to Britain and beyond. A knowledge of the general mobility of a species is essential in interpreting nearly all aspects of the monitoring results, but is especially important in relation to spatial distributions (Chapter 5).

In Table A1, three categories of mobility have been used, but no two species

Generations	Flight-period	Mobility	Male behaviour	Food plants
1	JJy	Sedentary	Patrol	*Nardus stricta*
1	JyA	Sedentary	Patrol	*Festuca* spp., *Brachypodium pinnatum*
1	JyA	Sedentary	Perch	*Festuca ovina, Ammophila aranaria*
1	JyA	Sedentary	pe/pa	Many grasses
1	J–S	Sedentary	Patrol	Many grasses
1	JyA	Sedentary	Patrol	*Dactylis glomerata* and other grasses
2*	MJ/AS	Sedentary	pe/pa	Many grasses
1	J–A	Sedentary	Patrol	*Eriophorum vaginatum*

*Each generation is split into groups which overlap, but often have distinct peaks of emergence.

are likely to be exactly comparable. The summary provided in the table is inevitably 'reasonable assumption'. In general, the information is based on the results of capture-mark-recapture studies. Such studies may show that recaptures of marked individuals are always close to the point of release, but if marked individuals leave the study area entirely there is little or no chance that they will be caught again. It is likely that the tendency is to underestimate mobility, especially of the common butterflies. For example, it seems rather contradictory that some of the butterflies that have expanded their ranges rapidly in recent years are generally considered to occur in sedentary populations.

FURTHER READING

Dennis, R.L.H. (ed) (1992) *Ecology of Butterflies in Britain*, Oxford University Press, Oxford.

Emmet, A.M. and Heath, J. (eds) (1989) *The Moths and Butterflies of Great Britain and Ireland*, Vol. 7, *The Butterflies*, Harley Books, Colchester.

Frohawk, F.W. (1934) *The Complete Book of British Butterflies*, Ward Lock, London.

Heath, J., Pollard, E. and Thomas, A.J. (1984) *Atlas of Butterflies in Britain and Ireland*, Viking, Harmondsworth.

Thomas, J.A. (1989) *Hamlyn Guide to the Butterflies of the British Isles*, 2nd edn, Hamlyn, London.

Thomas, J.A. and Lewington, R. (1991) *The Butterflies of Britain and Ireland*, Dorling Kindersley, London.

Vane-Wright, R.I. and Ackery, P.R. (eds) (1984) *The Biology of Butterflies*, Academic Press, London.

Appendix B
Sites and recorders in the Butterfly Monitoring Scheme

This list of sites includes all sites that have been recorded for four or more years and current sites recorded for two or more years. Data from a few of these sites have not been used in the calculation of collated 'all-sites' index values; this has been the case where there are sites of similar character in close proximity.

The status of the sites indicates ownership or body responsible for management: NNR, National Nature Reserve; RSPB, Royal Society for the Protection of Birds; NT, National Trust or National Trust for Scotland; NR, other type of nature reserve (mostly County Wildlife Trust); LA, local authority; FC, Forestry Commission; Private, owned by private individuals.

Current counties are listed, except in Scotland where Regions are used.

The recorders listed are the main recorders, or in a few cases the co-ordinators of recording. Many other people have made counts, sometimes for a season, sometimes as a substitute recorder for a few weeks and have contributed significantly to the scheme. We thank all recorders and apologize if there are any inaccuracies in the list or omissions from it.

Site	*Status*	*County*	*Recorder*
Alresford Farm	Private	Hampshire	J. Pain C.R. Cuthbert
Ampfield Wood	FC	Hampshire	G.C. Evans
Ariundle	NNR	Strathclyde	P. Duncan
Aston Rowan (2 routes)	NNR	Oxfordshire	M. Cox G. Ushaw D. Cooper R. Smith
Avon Gorge	NNR	Avon	R.V. Russell A. Robinson

Barnack Hills and Holes	NNR	Cambs.	M.E.S. Rooney D. Hughes D. Carstairs C. Gardiner
Batch Farm	Private	Glos.	M.K. Baker G.S. Baker
Ben Lawers	NT	Tayside	D.K. Mardon P.M. Batty B.D. Batty
Bevills Wood	FC	Cambs.	L. Farrell E. Pollard T.J. Yates N. Greatorex-Davis B.C. Eversham
Blean Woods	NNR	Kent	D.P. Maylam D.M.A. Rogers
Bovey Valley	NNR	Devon	P.A. Page R.G. Lamboll
Brockwells Farm	Private	Gwent	C. Titcombe E. Titcombe
Bure Marshes	NNR	Norfolk	M.J. Howat R. Southwood
Buttlers Hanging	NR	Bucks.	C. Hendry
Carnforth Marsh	NNR	Lancashire	R. Squires M. Robinson C. Wells
Castle Hill	NNR	Sussex	R. Leverton A.L. Bowley M.J. Emery
Castor Hanglands	NNR	Cambs.	P.W. Davis R. Harris
Chippenham Fen	NNR	Cambs.	M. Musgrave M. Wright
Church Place	FC	Hampshire	J. Gulliver
Church Wood	RSPB	Kent	M.F. Walter
Coedydd Maentwrog	NNR	Gwynedd	W.I. Jones
Coombes Valley	RSPB	Staffs	M. Waterhouse
Craig y Cilau	NNR	Powys	R.J. Haycock A.S. Ferguson

Derbyshire Dales	NNR	Derbyshire	A.J. Pritchard D. Gilbert C. Marsden B. Le Bas
Dyfi	NNR	Dyfed	R.B. Bovey P. Burnham M. Bailey
East Blean Woods	NR	Kent	M.A. Enfield J. McAllister
Ebbor Gorge	NNR	Somerset	T.L. Hodgson D.G. Thurlow P. Mountford
Folkestone Escarpment	Private	Kent	M.A. Enfield T.J. Yates J.N. Greatorex-Davis J. McAllister
Fontmell Down	NR	Dorset	W.G. Shreeves B. Yates Smith R. Wheeler O. Hooker J. Tubb
Foxholes	NR	Oxfordshire	M.P. Barnsley M.M. Cochrane
Gait Barrows	NNR	Lancashire	G.M. Barker A.C. Aldridge
Gibraltar Point	NNR	Lincolnshire	M.R. Curry R. Lambert C.J. Hawke K. Wilson
Gomm Valley	NR	Bucks.	D.H. Gantzel E.J. Ambrose
Hampstead Heath	LA	London	R.A. Softly
Ham St Woods (2 routes)	NNR	Kent	J.C. Maylam R. Petley-Jones D.P. Maylam
Hickling Broad	NNR	Norfolk	L. Dear C. Cogan
Holkham	NNR	Norfolk	D.A. Henshilwood R. Harold
Holme Dunes	NR	Norfolk	W. Boyd G.F. Hibberd

Holme Fen	NNR	Cambs.	P. Burnham R.N. Boston M. Rawlins A.L. Bowley
Insh Marshes	RSPB	Highland	R. Leavett Z. Bhatia
Inverpolly	NNR	Highland	A.G. Scott D.W. Duncan
Kingley Vale	NNR	Sussex	R.L. Williamson
Leigh Marshes	NNR	Essex	D. Cowan A.R. Mead S. Carter
Leighton Moss	RSPB	Lancashire	J. Wilson
Lindisfarne	NNR	Northumberland	P. Corkhill P.R. Davey
Loch Garten	RSPB	Highland	S. Taylor R. Thaxton
Loch Lomond	NNR	Strathclyde	J.M. Cameron J.H. Theaker
Luckett Wood	NR	Devon	C.L. Robbins R.T. Vulliamy C. Vulliamy
Lullington Heath	NNR	Sussex	A.L. Bowley M.J. Emery
Lydden	NR	Kent	I. Boyd J. McAllister
Martin Down	NNR	Hampshire	P. Toynton A. Knott R. Brunt
Monks Wood	NNR	Cambs.	E. Pollard M.L. Hall T.J. Yates
Moor Farm	NR	Lincolnshire	T.W. Bailey P. Roworth
Morris's Wood	Private	Sussex	W.R. Salkeld
Morrone Birkwood	NNR	Grampian	D. Batty P.M. Batty B. Marshall D. Marshall
Morton Lochs	NNR	Tayside	P.K. Kinnear

Mottistone Down	NR	Isle of Wight	A. Tutton P. Davies
Murlough	NNR	(N. Ireland)	J.A. Whatmough H.C. Thurgate
Nagshead	RSPB	Glos.	I.D. Bullock I. Proctor
Newborough Warren	NNR	Gwynedd (Anglesey)	V. Lane L.T. Colley
Newton Links	NT	Northumberland	D.L. Woodfall M.A. Freeman
Northward Hill	RSPB	Kent	A. Parker R.E. Scott
Oakley Wood	NR	Oxfordshire	R. Woodell A. Hyatt Williams
Old Winchester Hill	NNR	Hampshire	E.W. Baigent J.C. Bacon M.W. Finnemore B. Proctor
Oxwich	NNR	W. Glamorgan	D.O. Elias M.R. Hughes
Pewsey Down	NNR	Wiltshire	D. Painter K.R. Payne
Picket Wood	NR	Wiltshire	M. Fuller
Pitts Wood	FC	Hampshire	J.M. Noakes
Pollymore	NNR	Ross	A. Scott T.W. Henderson
Potton Wood	FC	Bedfordshire	I. Woiwod
Pynes Farmhouse	Private	Somerset	J. Ingram M.J. Ingram E. Sykes
Radipole Lake	RSPB	Dorset	D.T. Ireland A.R. Baker
Rostherne Mere	NNR	Cheshire	R.D. Fox M. Bailey M. Davey
Roudsea Wood	NNR	Lancashire	P.A. Singleton
St Cyrus	NNR	Grampian	D. Carstairs E.J. Cameron B. Lightfoot

St Margaret's Bay	NR	Kent	I. Hodgson P.J. Chantler A.J. Greenland
St Osyth	LA	Essex	R.W. Arthur B.C. Manning R.A.S. Marsh
Saltfleetby	NNR	Lincolnshire	T. Clifford G.P. Weaver A. Scott
Sands of Forvie	NNR	Grampian	R.B. Davis
Shabbington Wood	FC/NR	Oxfordshire	F.J. Hulbert B.M. Hulbert J. Collier
Skokholm	NR	Dyfed	J.R. Lawman R.S. Wolstenholme M. Betts S. Barclay
Skomer	NNR	Dyfed	R. Alexander S.J. Sutcliffe A.C. Sutcliffe
Smardale	NNR	Westmorland	R. Baines
Somerford Common (2 routes)	FC	Wiltshire	D. Stevens K.J. Grearson M. Tilzey L. Slade
South Stack (2 routes)	RSPB	Gwynedd (Anglesey)	I.D. Bullock A. Ferguson
Springhill Farm	Private	Kent	E. Pollard
Stour Wood	NNR	Essex	R. Leavett
Studland Heath	NNR	Dorset	J.R. Cox C.E. Ollivant
Swanage	NT	Dorset	J.A. Thomas
Tadnoll	NR	Dorset	J. Barker A. Bates Q. Palmer S. Frampton L. Swindall
Taynish	NNR	Strathclyde	B.D. Batty P.M. Batty J.B. Halliday
Tentsmuir Point	NNR	Tayside	P.K. Kinnear

Thorne Moors	NNR	S. Yorkshire	P. Rowarth
Upper Teesdale	NNR	Durham	I.H. Findlay
Vane Farm	RSPB	Tayside	R. Rowarth J. Stevenson
Walberswick	NNR	Suffolk	C.S. Waller J. Shackles
Wart Barrow	Private	Lancashire	I.R. Bonner M. Hay M. Hay B. Simpson
Waterperry Wood	NNR	Oxfordshire	K.M. Duparc F.E. Harvey G. Ushaw
Weeting Heath	NNR	Suffolk	R. Southwood M. Musgrave M. Wright
West Dean Woods	NR	Sussex	A. Williamson R.L. Williamson
Whitecross Green Wood	NR	Oxfordshire	R. Woodell J. Collier
Wicken Fen	NT	Cambs.	T.J. Bennet
Woodhurst	Private	Cambs.	V.P. Pollard
Woods Mill	NR	Sussex	M. Russell
Woodwalton Farm	Private	Cambs.	E. Pollard T.J. Yates M.G. Yates
Woodwalton Fen	NNR	Cambs.	R.I. Harold M.E. Massey A.L. Bowley
Wye	NNR	Kent	J.H. Duffield R.V. Russell
Wyre Forest	NNR	Worcestershire	M. Williams
Yarner Wood	NNR	Devon	D.A. Rogers D.G. Thurlow P.A. Page R.G. Lamboll
Ynis-hir	RSPB	Dyfed	R. Squires

Appendix C
Latin names of plants mentioned in text

Nomenclature follows Clapham *et al.* (1962)

Alder	*Alnus glutinosa*
Alder buckthorn	*Frangula alnus*
Ash	*Fraxinus excelsior*
Bilberry	*Vaccinium myrtilus*
Birch	*Betula* spp.
Birdsfoot trefoil	*Lotus corniculatus*
Bitter vetch	*Lathyrus montana*
Black medick	*Medicago lupulina*
Blackthorn	*Prunus spinosa*
Bluebell	*Endymion non-scriptus*
Bramble	*Rubus fruticosus* agg.
Bugle	*Ajuga reptans*
Clover	*Trifolium* spp.
Cocks foot	*Dactylis glomerata*
Common buckthorn	*Rhamnus catharticus*
Common cow-wheat	*Melampyrum pratense*
Common sorrel	*Rumex acetosa*
Couch grass	*Agropyron repens*
Creeping bent	*Agrostis stolonifera*
Creeping willow	*Salix repens*
Cross-leaved heath	*Erica tetralix*
Devil's bit scabious	*Succisa pratensis*
Dogwood	*Thelycrania sanguinea*
Elm	*Ulmus* spp.
European larch	*Larix decidua*
Germander speedwell	*Veronica chamaedrys*
Grass of Parnassus	*Parnassia palustris*
Greater knapweed	*Centaurea scabiosa*
Hawthorn	*Crataegus monogyna*
Hazel	*Corylus avellana*
Hedge garlic	*Alliaria petiolata*
Holly	*Ilex aquifolium*

Honeysuckle	*Lonicera periclymenum*
Hornbeam	*Carpinus betulus*
Horseshoe vetch	*Hippocrepis comosa*
Ivy	*Hedera helix*
Kidney vetch	*Anthyllis vulneraria*
Lady's smock	*Cardamine pratensis*
Marjoram	*Origanum vulgare*
Marram grass	*Ammophila arenaria*
Mat-grass	*Nardus stricta*
Meadowsweet	*Filipendula ulmaria*
Meadow vetchling	*Lathyrus pratensis*
Nettles	*Urtica dioica* and *U. urens*
Norway spruce	*Picea abies*
Oak	*Quercus robur* and *petraea*
Primrose	*Primula vulgaris*
Ragged robin	*Lychnis flos-cuculi*
Ragwort	*Senecio jacobaea*
Ribwort plantain	*Plantago lanceolata*
Rosebay willowherb	*Chamaenerion angustifolium*
Rushes	*Juncus* spp.
Sallow	*Salix atrocineria*
Sand sedge	*Carex arenaria*
Scots pine	*Pinus sylvestris*
Sedges	*Carex* spp.
Sheep's sorrel	*Rumex acetosella*
Spindle	*Euonymus europaeus*
Teasel	*Dipsacus fullonum*
Thistles	*Cirsium* spp.
Thyme	*Thymus drucei*
Tor grass	*Brachypodium pinnatum*
Tufted hair-grass	*Deschampsia cespitosa*
Valerian	*Valeriana officinalis*
Violet	*Viola* spp.
Wild strawberry	*Fragaria vesca*
Willow	*Salix* sp.
Wood false-brome	*Brachypodium sylvaticum*

REFERENCES

Abbot, C.H. (1951) A quantitative study of the migrations of the painted lady butterfly, *Vanessa cardui* L. *Ecology*, **32**, 155–71.

Arnold, R.A. (1980) *Ecological Studies of Six Endangered Butterflies (Lepidoptera, Lycaenidae): island biogeography, patch dynamics, and the design of habitat preserves*, University of California Publications, Entomology, no. 99, University of California Press.

Baker, R.R. (1968) Sun orientation during migration in some British butterflies. *Proceedings of the Royal Entomological Society of London (A)*, **43**, 89–95.

Baker, R.R. (1972a) Territorial behaviour of the nymphalid butterflies *Aglais urticae* (L.) and *Inachis io* (L.). *Journal of Animal Ecology*, **41**, 453–69.

Baker, R.R. (1972b) The geographical origin of the British spring individuals of the butterflies *Vanessa atalanta* and *V. cardui*. *Journal of Entomology (A)*, **46**, 185–96.

Baker, R.R. (1978) *The Evolutionary Ecology of Animal Migration*, Hodder and Stoughton, London.

Baker, R.R. (1984) The dilemma: when or how to go or stay, in *The Biology of Butterflies* (eds R.I. Vane-Wright and P.R. Ackery), Academic Press, London.

Barbour, D. (1986) Why are there so few butterflies in Liverpool? An answer. *Antenna*, **10**, 72–5.

Begon, M. (1979) *Investigating Animal Abundance*, Edward Arnold, London.

Beirne, B.P. (1955) Natural fluctuations in the abundance of British Lepidoptera. *Entomologist's Gazette*, **6**, 21–52.

Bibby, T.J. (1983) Oviposition by the brimstone *Gonepteryx rhamni* (L.) (Lepidoptera; Pieridae) in Monks Wood, Cambridgeshire in 1982. *Entomologist's Gazette*, **34**, 229–34.

Bink, F.A. (1985) Host plant preferences of some grass-feeding butterflies. *Proceedings of the 3rd Congress of European Lepidopterology*, **1982**, 23–9.

Bobbink, R. (1991) Effect of nutrient enrichment in Dutch chalk grassland. *Journal of Applied Ecology*, **28**, 28–41.

Brakefield, P.M. (1984) The ecological genetics of quantitative characters of *Maniola jurtina* and other butterflies, in *The Biology of Butterflies* (eds R.I. Vane-Wright and P.R. Ackery), Academic Press, London.

Brakefield, P.M. (1987) Geographical variation in, and temperature effects on, the phenology of *Maniola jurtina* and *Pyronia tithonus* (Lepidoptera, Satyrinae) in England and Wales. *Ecological Entomology*, **12**, 139–48.

Bretherton, R.F. and Chalmers-Hunt, J.M. (1981) The immigration of Lepidoptera to the British Isles in 1980, with an account of the invasion of the Painted Lady: *Cynthia cardui* L. Annexe III The Painted Lady (*Cynthia cardui* L.) in 1980. *Entomologist's Record and Journal of Variation* **93**, 103–11.

Bretherton, R.F. and Chalmers-Hunt, J.M. (1985) The immigration of Lepidoptera to the British Isles in 1981, 1982, 1983: a supplementary note. *Entomologist's Record and Journal of Variation*, **97**, 76–84.

BUTT (Butterflies Under Threat Team) (1986) *The Management of Chalk Grassland for Butterflies*, Nature Conservancy Council, Peterborough.

Chalmers-Hunt, J.M. and Owen, D.F. (1952) The history and status of *Pararge aegeria* (Lep. Satyridae) in Kent. *Entomologist*, **85**, 145–54.

Clapham, A.R., Tutin, T.G. and Warburg, E.F. (1962) *Flora of the British Isles*, 2nd edn, University Press, Cambridge.

Cochran, W.G. (1963) *Sampling Techniques*, Wiley, New York.

Cornell, H.V. and Lawton, J.H. (1992) Species interactions, local and regional processes, and limits to the richness of ecological communities: a theoretical perspective. *Journal of Animal Ecology*, **61**, 1–12.

Courtney, S.P. (1980) Studies on the biology of the butterflies *Anthocaris cardamines* (L.) and *Pieris napi* (L.) in relation to speciation in Pierinae. PhD thesis, University of Durham.

Craig, C.C. (1953) On the utilisation of marked specimens in estimating populations of flying insects. *Biometrika*, **40**, 170–77.

Davis, B.N.K. (1989) Habitat creation for butterflies on a landfill site. *Entomologist*, **108**, 109–22.

Davis, B.N.K. and Coppeard, R.P. (1990) Soil conditions and grassland establishment for amenity and wildlife on a restored landfill site, in *Biological Habitat Reconstruction* (ed. G.P. Buckley), Belhaven, London, pp.221–31.

Dempster, J.P. (1967) The control of *Pieris rapae* with DDT, I. The natural mortality of the young stages of *Pieris*. *Journal of Applied Ecology*, **4**, 485–500.

Dempster, J.P. (1975) *Animal Population Regulation*, Academic Press, London.

Dempster, J.P. (1983) The natural control of populations of butterflies and moths. *Biological Reviews*, **58**, 461–81.

Dempster, J.P. (1991) The role of intra-specific competition in determining insect abundance. *Antenna*, **15**, 105–9.

Dempster, J.P. and Pollard, E. (1981) Fluctuations in resource availability and insect populations. *Oecologia*, **50**, 412–16.

Dempster, J.P., Lakhani, K.H. and Coward, P.A. (1986) The use of chemical composition as a population marker in insects: a study of the Brimstone butterfly. *Ecological Entomology*, **11**, 51–65.

Dennis, R.L.H. (ed) (1977) *The British Butterflies: their origin and establishment*, Classey, Faringdon.

Dennis, R.L.H. (1982–83) Mate location strategies in the wall brown butterfly *Lasiommata megera* (L.) (Lepidoptera: Satyridae): wait or seek? *Entomologist's Record and Journal of Variation*, **94**, 209–14, **95**, 7–10.

Dennis, R.L.H. (1985a) Voltinism in British *Aglais urticae* (L.) (Lep. Nymphalidae): variation in space and time. *British Entomological and Natural History Society, Proceedings and Transactions*, **18**, 51–61.

Dennis, R.L.H. (1985b) *Polyommatus icarus* (Rottemburg) (Lepidoptera: Lycaenidae) on Brereton Heath in Cheshire: voltinism and switches in resource exploitation. *Entomologist's Gazette*, **36**, 175–9.

Dennis, R.L.H. (1992) *The Ecology of Butterflies in Britain*, Oxford University Press, Oxford.

Dennis, R.L.H. and Shreeve, T.G. (1989) Butterfly wing morphology variation in the British Isles: the influence of climate, behavioural posture and hostplant-habitat. *Biological Journal of the Linnaean Society*, **38**, 323–48.

Dennis, R.L.H. and Shreeve, T.G. (1991) Climatic change and the British butterfly fauna: opportunities and constraints. *Biological Conservation*, **55**, 1–16.

Department of the Environment (1988) *Possible Impacts of Climatic Change on the Natural Environment of the United Kingdom*. UK Department of the Environment, London.

Douwes, P. (1970) Size of, gain to and loss from a population of adult *Heodes virgaurea* (L.) (Lep. Lycaenidae), *Entomologica Scandinavica*, **1**, 263–87.

Downes, J.A. (1948) The history of the speckled wood butterfly (*Pararge aegeria*) in Scotland, with a discussion of recent changes of range of other British butterflies. *Journal of Animal Ecology*, **17**, 131–8.

Dunn, T.C. and Parrack, J.D. (1986) *The Moths and Butterflies of Northumberland and Durham*, Northern Naturalists Union, Durham.

Eberhardt, L.L. (1969) Population estimates from recapture frequencies *Journal of Wildlife Management*, **33**, 28–39.

Ehrlich, P.R., Breedlove, D.E., Brussard, P.F. and Sharp, M.A. (1972) Weather and the 'regulation' of subalpine populations. *Ecology*, **53**, 243–7.

Ekholm, S. (1975) Fluctuations in butterfly frequency in Central Nyland. *Notulae Entomologicae*, **40**, 65–80.

Elton, C. (1933) *The Ecology of Animals*, Methuen, London.

Emmet, A.M. and Heath, J. (eds) (1989) *The Moths and Butterflies of Great Britain and Ireland*, Vol. 7, *The Butterflies*, Harley Books, Colchester.

Evans, G.C. and Coombe, D.E. (1959) Hemispherical and woodland canopy photography and the light climate. *Journal of Ecology*, **47**, 103–13.

Fajer, E.D., Bowers, M.D. and Bazzaz, F.A. (1989) The effects of enriched carbon dioxide atmospheres on plant–insect herbivore interactions. *Science*, **243**, 1198–220.

Farrell, L. (1975) A survey of the status of the chequered skipper butterfly (*Carterocephalus palaemon*) (Pallas) (Lep. Hesperiidae) in Britain 1973–1974. *Entomologist's Gazette*, **26**, 148–9.

Feltwell, J. (1982) *The Large White Butterfly: the biology, biochemistry and physiology of* Pieris brassicae *(Linnaeus)*, Junk, The Hague.

Findlay, R., Young, M.R. and Findlay, J.A. (1983) Orientation behaviour in grayling butterflies: thermoregulation or crypsis? *Ecological Entomology*, **8**, 145–53.

Ford, E.B. (1945) *Butterflies*, Collins, London.

Ford, H.D. and Ford, E.B. (1930) Fluctuations in numbers and its influence on variation in *Melitaea aurinia* Rott. *Transactions of the Royal Entomological Society of London*, **78**, 345–52.

Frohawk, F.W. (1924) *The Natural History of British Butterflies* (2 volumes), Hutchinson, London.

Frohawk, F.W. (1934) *The Complete Book of British Butterflies*, Ward Lock, London.

Fry, R. and Lonsdale, D. (1991) *Habitat Conservation for Insects: a neglected green issue*, The Amateur Entomologists' Society, Middlesex.

Garcia-Barros, E. (1988) Delayed ovarian maturation in a butterfly *Hipparchia semele* as a possible response to summer drought. *Ecological Entomology*, **13**, 391–8.

Gilbert, F. and Owen, J. (1990) Size, shape, competition and community structure in hoverflies (Diptera: Syrphidae). *Journal of Animal Ecology*, **59**, 21–39.

Goddard, M.J. (1962) Broods of the speckled wood (*Pararge aegeria aegerides* Stgr.) (Lep. Satyridae). *Entomologist*, **95**, 289–307.

Goldsmith, F.B. (ed.) (1991) *Monitoring for Conservation and Ecology*, Chapman & Hall, London.

Greatorex-Davis, J.N., Sparks, T.H., Hall, M.L. and Marrs, R.H. (1992) The influence of shade on butterflies in rides of coniferised lowland woods in England and implications for conservation management. *Biological Conservation*, **63**, 31–41.

Greenwood, J.D. (1989) Bird population densities. *Nature*, **338**, 627–8.

Hall, M.L. (1981) *Butterfly Monitoring Scheme: instructions for independent recorders*, Institute of Terrestrial Ecology, Cambridge.

Hanski, I. (1982) Dynamics of regional distribution: the core and satellite species hypothesis. *Oikos*, **38**, 210–21.

Harrison, F. and Sterling, M.J. (1985) *Butterflies and Moths of Derbyshire*, Derbyshire Entomological Society, Derbyshire.

Heath, J., Pollard, E. and Thomas, J.A. (1984) *Atlas of Butterflies in Britain and Ireland*, Viking, Harmondsworth.

Higgins, L.G. and Reilly, N.D. (1970) *Butterflies of Britain and Europe*, Collins, London.

Holmes, J.W.O. (1978) A second brood of *Inachis io* (L.) (Lep. Nymphalidae) in 1976. *Entomologist's Gazette*, **29**, 42.

Hunt, R., Hand, D.W., Hannah, M.A. and Neal, A.M. (1991) Response to CO_2 enrichment in 27 herbaceous species. *Functional Ecology*, **5**, 410–21.

Jolly, G.M. (1965) Explicit estimates from capture–recapture data with both death and immigration – stochastic model. *Biometrika*, **52**, 225–47.

Kaye, W.J. (1900) Some diary notes on the season's collecting. *Entomologist's Record and Journal of Variation*, **12**, 233–35.

Larsen, T.B. (1976) The importance of migration to the butterfly faunas of the Lebanon, East Jordan and Egypt (Lepidoptera, Rhopalocera). *Notulae Entomologicae*, **56**, 73–93.

Lees, E. (1962) On the voltinism of *Coenonympha pamphilus* (L.) (Lep., Satyridae). *Entomologist*, **95**, 5–6.

Lees, E. (1965) Further observations on the voltinism of *Coenonympha pamphilus* (L.) (Lep., Satyridae) *Entomologist*, **98**, 43–5.

Lees, E. (1969) Voltinism of *Polyommatus icarus* Rott. (Lep., Lycaenidae) in Britain. *Entomologist*, **102**, 194–6.

Lees, E. and Archer, D.M. (1974) Ecology of *Pieris napi* (L.) (Lep., Pieridae) in Britain. *Entomologist's Gazette*, **25**, 231–7.

Lees, E. and Archer, D.M. (1980) Diapause in various populations of *Pieris napi* from different parts of the British Isles. *Journal of Research on the Lepidoptera*, **19**, 96–100.

Lipscombe, C.G. (1977) Second brood of *Inachis io*. *Entomologist's Record and Journal of Variation*, **89**, 18–19.

Lorimer, R.I. (1983) *The Lepidoptera of the Orkney Islands*, Classey Faringdon.

Manley, G.L. (1974) Central England temperatures: monthly means 1659–1973. *Quarterly Journal of the Royal Meteorological Society*, **100**, 389–405.

Marchant, J.H., Hudson, R., Carter, S.P. and Whittington, P. (1990) *Population Trends in British Breeding Birds*, British Trust for Ornithology, Tring.

McArdle, B.H., Gaston, K.J. and Lawton, J.H. (1990) Variation in the size of animal populations: patterns, problems and artefacts. *Journal of Animal Ecology*, **59**, 439–54.

McKay, H.V. (1991) Egg-laying requirements in woodland butterflies; brimstones (*Gonepteryx rhamni*) and alder buckthorn (*Frangula alnus*). *Journal of Applied Ecology*, **28**, 731–43.

Meteorological Office (1976–92) *Monthly Weather Reports*, HMSO, London.

Milne, A. (1957a) Theories of natural control of insect populations. *Cold Spring Harbour Symposium on Quantitative Biology*, **22**, 253–71.

Milne, A. (1957b) The natural control of insect populations. *Canadian Entomologist*, **89**, 193–213.

Milne, A. (1962) On a theory of natural control of insect populations. *Journal of Theoretical Biology*, **3**, 19–50.

Moore, N.W. (1975) Butterfly transects in a linear habitat, 1964–73. *Entomologist's Gazette*, **26**, 71–8.

Morisita, M. (1959) Measuring the dispersion of individuals and analysis of the distribution patterns. *Memoir of the Faculty of Science of Kyushu University, Series E, Biology*, **2**, 215–35.

Morton, A.C. (1984) The effects of marking and handling on recapture frequencies of butterflies, in *The Biology of Butterflies* (eds R.I. Vane-Wright and P.R. Ackery), Academic Press, London.

Moss, D. (1985) Some statistical checks on the BTO Common Birds Census Index – 20 years on, in *Bird Census and Atlas Studies* (eds K. Taylor, R.J. Fuller and P.C. Lack), British Trust for Ornithology, Tring, pp. 175–9.

Moss, D. and Pollard, E. (1993) Calculation of collated indices of abundance of butterflies based on monitored sites. *Ecological Entomology* (in press).

Nylin, S. (1989) Effects of changing photoperiods in the life-cycle of the comma butterfly *Polygonia c-album* (Nymphalidae). *Ecological Entomology*, **14**, 209–18.

Oates, M. and Warren, M.S. (1990) *A Review of Butterfly Introductions in Britain and Ireland*. Nature Conservancy Council, Peterborough.

Owen, D.F. (1975) Estimating the abundance and diversity of butterflies. *Biological Conservation*, **8**, 173–83.

Pavlicek-van Beek, T., Ovaa, A.H. and van der Made, J.G. (eds) (1992) *Future of Butterflies in Europe*, Agricultural University, Wageningen.

Pimm, S.L., Jones, H.L. and Diamond, J. (1988) On the risk of extinction. *American Naturalist*, **132**, 757–85.

Pivnick, K.A. and McNeil, J.N. (1987) Diel patterns of activity of *Thymelicus lineola* adults (Lepidoptera: Hesperiidae) in relation to weather. *Ecological Entomology*, **12**, 197–207.

Plant, C.W. (1987) *Atlas of Butterflies of the London Area*, London Natural History Society, London.

Pollard, E. (1971) Hedges VI. Habitat diversity and crop pests: a study of *Brevicoryne brassicae* and its syrphid predators. *Journal of Applied Ecology*, **8**, 751–80.

Pollard, E. (1977) A method for assessing changes in the abundance of butterflies. *Biological Conservation*, **12**, 115–34.

Pollard, E. (1979a) Population ecology and change in range of the white admiral butterfly *Ladoga camilla* L. in England. *Ecological Entomology*, **4**, 61–74.

Pollard, E. (1979b) A national scheme for monitoring the abundance of butterflies: the first three years. *British Entomological and Natural History Society, Proceedings and Transactions*, **12**, 77–90.

Pollard, E. (1981) Aspects of the ecology of the meadow brown butterfly *Maniola jurtina* (L.). *Entomologist's Gazette*, **32**, 67–74.

Pollard, E. (1982a) Observations on the migratory behaviour of the painted lady butterfly, *Vanessa cardui* (L.) (Lepidoptera, Nymphalidae). *Entomologist's Gazette*, **33**, 99–103.

Pollard, E. (1982b) Monitoring the abundance of butterflies in relation to the management of a nature reserve. *Biological Conservation*, **24**, 317–28.

Pollard, E. (1984) Fluctuations in the abundance of butterflies, 1976–82. *Ecological Entomology*, **9**, 179–88.

Pollard, E. (1985) Larvae of *Celastrina argiolus* (L.) (Lepidoptera: Lycaenidae) on male holly bushes. *Entomologist's Gazette*, **36**, 3.

Pollard, E. (1988) Temperature, rainfall and butterfly numbers. *Journal of Applied Ecology*, **25**, 819–28.

Pollard, E. (1991a) Synchrony of population fluctuations: the dominant influence of widespread factors on local butterfly populations. *Oikos*, **60**, 7–10.

Pollard, E. (1991b) Changes in the flight period of the hedge brown butterfly *Pyronia tithonus* during range expansion. *Journal of Animal Ecology*, **60**, 737–48.

Pollard, E. (1992) Monitoring populations of a butterfly during a period of range expansion, in *Biological Recording of Changes in British Wildlife* (ed. P.T. Harding), HMSO, London.

Pollard, E. and Hall, M.L. (1980) Possible movement of *Gonepteryx rhamni* (L.) (Lepidoptera: Pieridae) between hibernating and breeding areas. *Entomologist's Gazette*, **31**, 217–20.

Pollard, E. and Lakhani, K.H. (1985) Butterfly Monitoring Scheme: effects of weather on abundance. *Institute of Terrestrial Ecology, Annual Report, 1984*, 54–6.

Pollard, E. and Leverton, R. (1991) Monitoring butterfly numbers. Case study: Castle Hill National Nature Reserve, in *Monitoring for Conservation and Ecology*, (ed. B. Goldsmith), Chapman & Hall, London.

Pollard, E. and Yates, T.J. (1992) The extinction and foundation of local butterfly populations in relation to population variability and other factors. *Ecological Entomology*, **17**, 249–54.

Pollard, E. and Yates, T.J. (1993) Population fluctuations of the holly blue butterfly *Celestrina argiolus* (L.), *Entomologist's Gazette* (in press).

Pollard, E., Elias, D.O., Skelton, M.J. and Thomas, J.A. (1975) A method of assessing the abundance of butterflies in Monks Wood National Nature Reserve in 1973. *Entomologist's Gazette*, **26**, 79–88.

Pollard, E., Hall, M.L. and Bibby, T.J. (1984) The clouded yellow butterfly migration of 1983. *Entomologist's Gazette*, **35**, 227–34.

Pollard, E., Hall, M.L. and Bibby, T.J. (1986) *Monitoring the Abundance of Butterflies 1976–1985*, Nature Conservancy Council, Peterborough.

Porter, K. (1981) The population dynamics of small colonies of the butterfly *Euphydryas aurinia*. D. Phil. thesis, University of Oxford.

Pratt, C.R. (1981) *A History of the Butterflies and Moths of Sussex*, Borough Council, Brighton.

Pratt, C.R. (1986–87) A history and investigation into the fluctuations of *Polygonia c-album* L. : the comma butterfly. *Entomologist's Record and Journal of Variation*, **98**, 197–203, 244–50, **99**, 21–7, 69–80.

Pullin, A.S. (1986) Effect of photoperiod and temperature on the life-cycle of different populations of the peacock butterfly *Inachis io*. *Entomologia Experimentalis et Applicata*, **41**, 237–42.

Pullin, A.S. (1987) Adult feeding time, lipid accumulation, and overwintering in *Aglais urticae* and *Inachis io* (Lepidoptera: Nymphalidae). *Journal of Zoology, London*, **211**, 631–41.

Rose, D.J.W., Page, W.W., Dewhurst, C.F., Riley, J.R., Reynolds, D.R., Pedgely, D.E. and Tucker, M.R. (1985) Downwind migration in the African armyworm moth, *Spodoptera exempta*, studied by mark-and-capture and by radar. *Ecological Entomology*, **10**, 299–313.

Scali, V. (1971) Imaginal diapause and gonadal maturation of *Maniola jurtina* (Lepidoptera: Satyridae) from Tuscany. *Journal of Animal Ecology*, **40**, 467–72.

Scott, J.A. (1977) Competitive exclusion due to mate searching behaviour, male–female emergence lags and fluctuations in number of progeny in model invertebrate populations. *Journal of Animal Ecology*, **46**, 909–24.

Shreeve, T.G. (1984) Habitat selection, mate location and microclimate constraints on the activity of the speckled wood butterfly *Pararge aegeria*. *Oikos*, **42**, 371–7.

Shreeve, T.G. (1989) The extended flight-period of *Maniola jurtina* on chalk downland: seasonal changes of the adult phenotype and evidence for a population of mixed origins. *Entomologist*, **108**, 202–15.

Singer, M.C. (1982) Sexual selection for small size in male butterflies. *American Naturalist*, **119**, 440–3.

Singer, M.C. and Wedlake, P. (1981) Capture does affect probability of recapture in a butterfly species. *Ecological Entomology*, **6**, 215–16.

South, R. (1906) *The Butterflies of the British Isles*, Warne, London.

Steele, R.C. and Welch, R.C. (1973) *Monks Wood: a nature reserve record*, The Nature Conservancy, Huntingdon.

Sterling, P.H. and Hambler, C. (1988) *Coppicing for Conservation: do Hazel Communities Benefit?*, Nature Conservancy Council, Peterborough.

Sutton, S.L. and Beaumont, H.E. (1989) *Butterflies and Moths of Yorkshire*, Yorkshire Naturalist's Union, Yorkshire.

Taylor, L.R. (1977) Aphid forecasting and the Rothamsted Insect Survey. R.A.S.E. Research Medal Paper. *Journal of the Royal Agricultural Society of England*, **138**, 75–97.

Taylor, L.R. (1979) *Agriculture, Urbanisation and Conservation*. Hertfordshire and Middlesex Trust for Nature Conservation, **49**, 1–3.

Taylor, L.R. (1986) Synoptic dynamics, migration and the Rothamsted Insect Survey, Presidential Address. *Journal of Animal Ecology*, **55**, 1–38.

Thomas, C.D. (1991) Spatial and temporal variability in a butterfly population. *Oecologia*, **87**, 577–80.

Thomas, J.A. (1974) Ecological studies of hairstreak butterflies. PhD thesis, University of Leicester.

Thomas, J.A. (1980) Why did the large blue become extinct in Britain? *Oryx*, **15**, 243–7.

Thomas, J.A. (1983a) A quick method of estimating butterfly numbers during surveys. *Biological Conservation*, **27**, 195–211.

Thomas, J.A. (1983b) The ecology and conservation of *Lysandra bellargus* (Lepidoptera: Lycaenidae) in Britain. *Journal of Applied Ecology*, **20**, 59–83.

Thomas, J.A. (1984) The conservation of butterflies in temperate countries: past efforts and lessons for the future, in *The Biology of Butterflies* (eds R.I. Vane-Wright and P.R. Ackery), Academic Press, London.

Thomas, J.A. (1986) *Butterflies of the British Isles*, Country Life Books, London.

Thomas, J.A. (1990) The conservation of Adonis blue and Lulworth skipper butterflies – two sides of the same coin, in *Calcareous Grasslands: Ecology and management* (eds S.H. Hillier, D.W.H. Walton and D.A. Wells), Bluntisham Books, Huntingdon.

Thomas, J.A. (1991) Rare species conservation: butterfly case studies, in *The Scientific Management of Temperate Communities for Conservation* (eds I.F. Spellerburg, F.B. Goldsmith and M.G. Morris), Blackwell, Oxford.

Thomas, J.A. and Lewington, R. (1991) *The Butterflies of Britain and Ireland*, Dorling Kindersley, London.

Thomas, J.A. and Merrett, P. (1980) Observations of butterflies in the Purbeck Hills in 1976 and 1977. *Proceedings of the Dorset Natural History and Archaeological Society*, **99**, 112–19.

Thomas, J.A. and Webb, N.R.C. (1984) *Butterflies of Dorset*, Dorset Natural History and Archaeological Society, Dorchester.

Thompson, J.A. (1952) Butterflies in the coastal region of North Wales. *Entomologist's Record and Journal of Variation*, **64**, 161–6.

Thomson, G. (1980) *The Butterflies of Scotland*, Croom Helm, London.

Tilden, J.W. (1962) General characteristics of the movements of *Vanessa cardui* (L.) *Journal of Research on the Lepidoptera*, **1**, 43–9.

Turner, J.R.G. (1986) Why are there so few butterflies in Liverpool? Homage to Alfred Russel Wallace. *Antenna*, **10**, 18–24.

Turner, J.R.G., Gatehouse, C.M. and Corey, C.A. (1987) Does solar energy control organic diversity? Butterflies, moths and the British climate. *Oikos*, **48**, 195–205.

Urquhart, F.A. (1976) Found at last: the monarch's winter home. *National Geographical Magazine*, **150**, 161–73.

Urquhart, F.A. and Urquhart, N.R. (1978) Autumn migration routes of the eastern population of the monarch butterfly (*Danaus p. plexippus*) in North America to overwintering sites in the Neovolcanic Plateau of Mexico. *Canadian Journal of Zoology*, **56**, 1754–64.

van Swaay, C.A.M. (1990) An assessment of the changes in butterfly abundance in the Netherlands during the 20th century. *Biological Conservation*, **52**, 287–302.

Vepsalainen, K. (1968) The immigration of *Pieris brassicae* (L.) (Lep., Pieridae) into Finland in 1966, with a general discussion of insect migration. *Annales Entomologica Fennici*, **34**, 223–43.

Walker, T.J. (1991) Butterfly migration from and to peninsular Florida. *Ecological Entomology*, **16**, 241–52.

Walker, T.J. and Riordan, A.J. (1981) Butterfly migration, are synoptic-scale winds important? *Ecological Entomology*, **6**, 433–40.

Warren, M.S. (1981) The ecology of the wood white butterfly *Leptidea sinapis* L. (Lepidoptera, Pieridae) PhD thesis, University of Cambridge.

Warren, M.S. (1984) The biology and status of the wood white butterfly, *Leptidea sinapis* (L.) (Lepidoptera: Pieridae), in the British Isles. *Entomologist's Gazette*, **35**, 207–23.

Warren, M.S. (1985) The influence of shade on butterfly numbers in woodland rides, with special reference to the wood white *Leptidea sinapis*. *Biological Conservation*, **33**, 147–64.

Warren, M.S. (1987a) The ecology and conservation of the heath fritillary butterfly, *Mellicta athalia*. I. Host selection and phenology. *Journal of Applied Ecology*, **24**, 467–82.

Warren, M.S. (1987b) The ecology and conservation of the heath fritillary butterfly, *Mellicta athalia*. II. Adult population structure and mobility. *Journal of Applied Ecology*, **24**, 483–98.

Warren, M.S. (1987c) The ecology and conservation of the heath fritillary butterfly, *Mellicta athalia*. III. Population dynamics and the effects of habitat management. *Journal of Applied Ecology*, **24**, 499–513.

Warren, M.S. (1991) The successful conservation of an endangered species, the heath fritillary butterfly, *Mellicta athalia*, in Britain. *Biological Conservation*, **55**, 37–56.

Warren, M.S. (1993) A review of butterfly conservation in central southern Britain. I. Protection, evaluation and extinction in prime sites. *Biological Conservation*, in press.

Warren, M.S., Pollard, E. and Bibby, T.J. (1986) Annual and long-term changes in a population of the wood white butterfly *Leptidea sinapis*. *Journal of Animal Ecology*, **55**, 707–19.

Warren, M.S., Thomas, C.D. and Thomas, J.A. (1984) The status of the heath fritillary *Mellicta athalia* Rott. in Britain. *Biological Conservation*, **29**, 287–305.

Welch, R.C. (1989) Dutch elm disease in Huntingdonshire and its effect upon the associated insect fauna, in *40 Years of Change in the County* (eds T.C.E. Wells, J.H. Cole and P.E.G. Walker), Huntingdonshire Fauna and Flora Society, Huntingdon.

Wickman, P.-O., Wiklund, C. and Karlsson, B. (1990) Comparative phenology of four satyrine butterflies inhabiting dry grasslands in Sweden. *Holarctic Ecology*, **13**, 238–46.

Wiklund, C. (1977a) Oviposition, feeding and spatial separation of breeding and foraging habits in a population of *Leptidea sinapis* (Lepidoptera). *Oikos*, **28**, 56–68.

Wiklund, C. and Fagerström, T. (1977b) Why do males emerge before females? *Oecologia*, **31**, 153–8.

Wiklund, C. (1984) Egg-laying patterns in butterflies in relation to their phenology and visual apparency and abundance of their hosts. *Oecologia*, **63**, 23–9.

Williams, C.B. (1930) *The Migration of Butterflies*, Oliver and Boyd, Edinburgh.

Williams, C.B. (1958) *Insect Migration*, Collins, London.

Wilson, J. (1991) Population changes in three butterfly species in north Lancashire. *The Naturalist*, **116**, 95–8.

Winstanley, D., Spencer, R. and Williamson, K. (1974) Where have all the whitethroats gone? *Bird Study*, **21**, 1–14.

Winter, P.Q. (1981) The painted lady (*Cynthia cardui* L.) in numbers at light and other notes on its occurrence near Filey, Yorkshire in 1980. *Entomologist's Record and Journal of Variation*, **93**, 303–4.

Woiwod, I. and Dancy, K.J. (1987) Synoptic monitoring of migrant insect pests in Great Britain and Western Europe. VII Annual population fluctuations of macrolepidoptera over Great Britain for 17 years. *Rothamsted Experimental Station. Annual Report for 1986, Part 2*, 237–64.

Zonneveld, C. (1991) Estimating death rates from transect counts. *Ecological Entomology*, **16**, 115–21.

Site Index

The sites indexed are those mentioned in the text which have contributed data to the Butterfly Monitoring Scheme. Not all sites in the scheme are mentioned in the text, but all (listed in Appendix 2) have contributed to the synoptic results presented in the book.

Species Index

Subject Index